Local Responses to Mine Closure in South Africa

This book investigates mine closure and local responses in South Africa, linking dependencies and social disruption.

Mine closure presents a major challenge to the mining industry and government policymakers globally, but particularly in the Global South. South Africa is experiencing notable numbers of mine closures, and this book explores the notion of social disruption, a concept often applied to describe the effects of mine growth on communities but often neglecting the impact of mine closures. The book begins with three theoretical chapters that discuss theory, closure cost frameworks and policy development in South Africa. It uses evolutionary governance theory to show how mining creates dependencies and how mining growth often blinds communities and governments to the likelihood of closure. Too easily, mining goes ahead with no concern for the possibility, or indeed inevitability, of eventual closure and how mining communities will cope. These impacts are showcased through eight place-based case studies from across South Africa, one focusing on mine workers, to demonstrate that mine closure causes significant social disruption.

This book will be of interest to students and scholars researching the social impacts of mining and the extractive industries, social geography and sustainable development, as well as policymakers and practitioners working with mine closure and social impact assessments.

Sethulego Matebesi is Associate Professor and Academic Head of the Department of Sociology at the University of the Free State (UFS), South Africa. His previous publications include *Civil Strife Against Local Governance: Dynamics of Community Protests in South Africa* (2017) and *Social Licensing and Mining in South Africa* (Routledge, 2020).

Lochner Marais is Professor of Development Studies at the Centre for Development Support at the University of the Free State (UFS), South Africa, and an executive editor of the newly established journal *Research Directions: Mine Closure and Transitions*. He is also an honorary professor at the Sustainable Minerals Institute, University of Queensland, Australia. He has published *The Social Impacts of Mine Closure in South Africa: Housing Policy and Place Attachment* (Routledge, 2022).

Verna Nel is Professor of Urban and Regional Planning at the University of the Free State, South Africa. She is also qualified as a town and regional planner at Wits University, South Africa.

Routledge Studies of the Extractive Industries and Sustainable Development

The Impact of Mining Lifecycles in Mongolia and Kyrgyzstan
Political, Social, Environmental and Cultural Contexts
Edited by Troy Sternberg, Kemel Toktomushev and Byambabaatar Ichinkhorloo

Oil and National Identity in the Kurdistan Region of Iraq
Conflicts at the Frontier of Petro-Capitalism
Alessandro Tinti

The Anthropology of Resource Extraction
Edited by Lorenzo D'Angelo and Robert Jan Pijpers

Andean States and the Resource Curse
Institutional Change in Extractive Economies
Edited by Gerardo Damonte and Bettina Schorr

Stakeholders, Sustainable Development Policies and the Coal Mining Industry
Perspectives from Europe and the Commonwealth of Independent States
Izabela Jonek-Kowalska, Radosław Wolniak, Oksana A. Marinina and Tatyana V. Ponomarenko

The Social Impacts of Mine Closure in South Africa
Housing Policy and Place Attachment
Lochner Marais

Local Responses to Mine Closure in South Africa
Dependencies and Social Disruption
Edited by Sethulego Matebesi, Lochner Marais and Verna Nel

For more information about this series, please visit: www.routledge.com/Routledge-Studies-of-the-Extractive-Industries-and-Sustainable-Development/book-series/REISD

Local Responses to Mine Closure in South Africa

Dependencies and Social Disruption

Edited by Sethulego Matebesi, Lochner Marais and Verna Nel

Routledge
Taylor & Francis Group
LONDON AND NEW YORK

earthscan
from Routledge

First published 2024
by Routledge
4 Park Square, Milton Park, Abingdon, Oxon OX14 4RN

and by Routledge
605 Third Avenue, New York, NY 10158

Routledge is an imprint of the Taylor & Francis Group, an informa business

British Library Cataloguing-in-Publication Data
A catalogue record for this book is available from the British Library

ISBN: 978-1-032-51043-9 (hbk)
ISBN: 978-1-032-51658-5 (pbk)
ISBN: 978-1-003-40332-6 (ebk)

DOI: 10.324/9781003403326

Typeset in Times New Roman
by Apex CoVantage, LLC

Contents

Tables

Figures

Acronyms

CEO	Chief Executive Officer
CO_2,	Carbon dioxide
CPA	Communal Property Association
CSR	Corporate Social Responsibility
DMRE	Department of Minerals, Resources and Energy
DPE	Department of Public Enterprises
FSGDC	Free State Goldfields Development Centre
GPN	Global Protection Network
GDP	Gross Domestic Product
GVA	Gross Value Add
HIV/AIDS	Human immunodeficiency virus/acquired immunodeficiency syndrome
ICT	Information and Communication Technology
IDP	Integrated Development Plan
IEEFA	Institute for Energy Economics and Financial Analysis
IRP	Integrated Resource Plan
TKLA	Traditional and Khoi-San Leadership
NEMA	National Environmental Management Act
MACUA	Mining Affected Communities United in Action
MMIC	Matjhabeng Marketing and Investment Company
MPRDA	Minerals, Petroleum and Resources Development Act
NGO	Non-Governmental Organisation (NGO)
OVC	Orphans and Vulnerable Children
PCC	Presidential Climate Commission
PSJV	Pooling and Sharing Joint Venture
R&D	Research and Development
RBNDT	Royal Bafokeng Nation Development Trust
RDP	Reconstruction and Development Programmes
RMC	Richtersveld Mining Company
SADC	Southern African Development Community
SASSA	South African Social Security Agency
SEDA	Small Enterprise Development Agency
SEZ	Special Economic Zone
SDF	Spatial Development Framework

SIA	Social Impact Assessment
SLP	Social and Labour Plan
SMME	Small Medium and Micro Enterprise
UK	United Kingdom
US	United States

Contributors

Boitumlo Joy Alec is a post-graduate student at the Centre for Development Support, University of the Free State, Bloemfontein, South Africa.

Derick Blaauw is Professor of Economics, School of Economic Sciences, North-West University, Potchefstroom, South Africa.

Lucius Botes is Professor and Managing Director: PhDs, North West University, Vanderbijlpark, South Africa.

Maléne Campbell is Professor Emeritus, Department of Urban and Regional Planning, University of the Free State, Bloemfontein, South Africa.

Jan Cloete is Researcher, Centre for Development Support, University of the Free State, Bloemfontein, South Africa.

Cathy M Dzerefos, Lecturer, Tshwane University of Technology, Pretoria, South Africa.

Job Gbadegesin is Research Associate, Centre for Development Support, University of the Free State, Bloemfontein, South Africa.

Oupa Kale is a post-graduate student at the Centre for Development Support, University of the Free State, Bloemfontein, South Africa.

Samkelisiwe Khanyile is Researcher, Gauteng City-Region Observatory, Johannesburg, South Africa.

Martina Kotzé is Professor, Department of Industrial Psychology, University of the Free State, Bloemfontein, South Africa.

Margaret Kusambiza-Kiingi is a post-graduate student at the Centre for Development Support, University of the Free State, Bloemfontein, South Africa.

Dolphin Mabale is Lecturer, Department of Human Sciences, University of Venda, Thohoyandou, South Africa.

Mmboneni Magadzu is a post-graduate student at the Centre for Development Support, University of the Free State, Bloemfontein, South Africa.

Ngaka Mosiane is Researcher, Gauteng City-Region Observatory, Johannesburg, South Africa.

Zolile Mqotyana is a post-graduate student at the Centre for Development Support, University of the Free State, Bloemfontein, South Africa.

John Ntema is Associate Professor, Department of Development Studies, Unisa, South Africa.

Anmar Pretorius is Professor of Economics, School of Economic Sciences, North-West University, Potchefstroom, South Africa.

Kentse Sesele is a post-doctoral fellow, Centre for Development Support, University of the Free State, Bloemfontein, South Africa.

Thomas Stewart is Lecturer, Department of Urban and Regional Planning, University of the Free State, Bloemfontein, South Africa.

Phia van der Watt is Research Associate, Centre for Development Support, University of the Free State, Bloemfontein, South Africa.

David van Wyk, Bench Marks Foundation, Johannesburg, South Africa.

Anita Venter is Researcher, Centre for Development Support, University of the Free State, Bloemfontein, South Africa.

1 Understanding mine closure

Global and national trends

Lochner Marais, Sethulego Matebesi and Verna Nel

1.1 A future beyond mining

In September 2021, the 40-year-old general manager of a mine in South Africa spoke to Lochner Marais about his personal plan for a post-mining future. He said the mine is going to close in 2026, and, although he is a qualified engineer, he had enrolled for an MBA degree. A business qualification could help him find a new livelihood. He was unlikely to resign before closure (we suspected bonuses were at stake), but he was sensibly preparing for life beyond mine closure. Apart from his concern for himself, he was also concerned about the mine's suppliers, many of whom have only one client, the mine. When the mine closes, their businesses will close as well. Even though he has repeatedly told them about the impending closure, they continue to supply only the mine as they do not want to believe the closure date is real.

This anecdote epitomises the syndrome that this book describes in chapter after chapter: an unwillingness to think about closure and its consequences. It illustrates a form of cognitive lock-in that is common in mining environments and extends well beyond suppliers. In town after town, people employed by the mines and those not directly involved in mining yet also at risk from its fickleness stubbornly ignore closure signals. It encapsulates other themes running through this book. Mining creates dependencies. It supports non-mining livelihoods, but these are vulnerable. Mining booms blind people to the inevitability of closure. Unlike the mines, most of which are fairly well prepared, communities and governance structures make little or no preparation for closure. The result of this failure to plan early in a mine's life cycle for a future beyond mining is social disruption and trauma, to which many of the case studies in this book vividly bear witness.

1.2 The book's contributions

Our book fills a specific research gap. We believe that research has overemphasised the social consequences when a mine opens and during its working life and failed to give sufficient attention to the consequences of closure.

Over the past 20 years, the mine closure literature has grown, but several concerns remain, particularly that it is fragmented and dominated by environmental

DOI: 10.324/9781003403326-1

aspects (Bainton & Holcombe, 2018; Ndaguba & Marais, 2023). Not many books have been written on the social aspects of mine closure. To date, those we know of are Neil et al. (1992), Chaloping-March (2017), Marais (2023) and Van Assche et al. (2024, in press). Besides being fragmentary, the literature is dominated by single case studies. Indeed, we ourselves use case studies in this book, but we argue that this collection invites comparison and helps to develop a comprehensive view. We suggest more comparative work like this, both within and across countries, should be done. Another concern about the literature is that it is descriptive and atheoretical. In this book, we use theory to illuminate the material the descriptive case studies provide. We selected two theoretical frameworks to understand mine closure: the concept of social disruption (often used to explain the social impacts of large mining investments) and the theory of dependencies as outlined in evolutionary governance theory (see Chapter 2). Yet another concern is that research on environmental and social aspects of mine closure remains mainly in the hands of private firms (Bainton & Holcombe, 2018). This book is an attempt to broaden the knowledge base, extending the literature.

A further concern is that research on the social aspects of mining over the past 50 years has sought to find ways to ensure that mining will have local benefits. Some terms that are common in this approach are 'social licence to operate', 'responsible mining', 'corporate social responsibility' and 'collaborative planning'. One of the main problems with this approach has been that it focuses on the value of mining when a mine opens. This focus needs to shift to how that value might be negated by closure and what value might be garnered from closure. Although maximising local benefit during the mine's life cycle is important, it can create dependencies, as illustrated by the precarious position of the suppliers the general manager was worried about.

Historically, mining legislation did not pay much attention to closure. Authorities provided mining licences but did not regulate closure. Many mines terminated activities and abandoned their operations when they became unprofitable. South Africa's 6,000 abandoned mines are a case in point (Watson & Olalde, 2019). Over the past 50 years, global and South African mine closure legislation has become more stringent (see Chapter 3). Historically, companies simply abandoned mines, and local communities were left to live with this reality. Vivoda et al. (2019) created a framework for distinguishing the social aspects of mine closure. The framework, shown in Table 1.1, has 13 social aspects and 27 indicators.

In the final chapter of the book, we revisit this conceptual framework. We find it a useful list of aspects to be considered during a mine's decline and closure, but it has some shortcomings. For example, it does not list individual and community trauma as a specific aspect. We feel this needs emphasis. Van Assche et al. (2021) have shown how this trauma remains with communities and inhibits the development of alternative economic possibilities. Closure often has intergenerational consequences, which the framework does not include. As the framework does not include causes, it does not show that many of the social problems at closure spring from decisions taken during booms. And as the framework was designed for Australia, it does not consider aspects specific to Africa. Many African governments view mining as a step towards development and modernity. The mine closure

Table 1.1 Social aspects of mine closure

Social aspect	*Indicative elements*
Economic	Local economic activity (diversity and dependence) Household income Local living standards
Business	Local business development opportunities in the mining Business opportunities in other sections
Employment	Local employment opportunities in the mining Local employment opportunities in other sectors Local employment stability/volatility
Security, education and training infrastructure	Social order and safety (e.g. tensions, crime and violence) Local skills development in mining and other sectors Access to quality education and training Local transport (e.g. public buses, roads and airports) Critical infrastructure (food supply, power supply, water supply, telecommunications)
Amenity	Local aesthetic and recreational resources (e.g. heritage sites, parks, recreation areas, communal areas) Local culture, arts and sports (including facilities)
Livelihoods	Local livelihoods (e.g. access to land, food, water and shelter that affects livelihoods)
Land	Local land access, ownership and use Recognition of traditional, customary ownership
Housing and health	Local housing quality, availability and affordability Community health and wellbeing Access to quality health and social services
Environment	Environmental aspects that affect social conditions (e.g. quality of air, water, land, ecosystem)
Demography	Local population dynamics (e.g. growth/decline, migration, ageing, gender balance)
Participation	Stakeholder participation in closure planning and closure and post-closure processes (including decision-making)
Inclusion	Inclusive stakeholder engagement, including vulnerable and otherwise marginalised groups (e.g. Indigenous peoples, women, ethnic minorities, disabled, elderly, young) in closure planning and closure and post-closure processes (including decision-making)
Social (general)	General socio-economic considerations Social considerations in financial assurance mechanisms

Source: Vivoda et al. (2019, p. 8)

disrupts this view. In the final chapter, we offer our expanded version of Vivoda et al.'s framework, using findings from the case studies, taking into account aspects specific to mine closure in South Africa, and broadening the debate by linking the aspects to social disruption and dependency.

1.3 The mining town and closure

The mine closure takes place in specific spaces. Although many mines are in remote areas with nearby settlements, most of those in South Africa are near urban areas (often small towns or secondary cities). Remote mining operations typically need help attracting skilled people to their sites. In response, mining companies developed company towns to reduce living costs and provide a comfortable living environment (Marais et al., 2018). The companies managed these towns to ensure that they served as an attraction. In many cases, the employees living in the company town did not have to pay rent and received water and electricity services at a discount. However, mining companies that did this have been criticised for, among other things, usurping public functions and playing a dominant role in public affairs (Obeng-Odoom, 2009). 'Normalisation', in other words, changing company towns from private to public management, thus became popular. In Canada, it had become the norm by the 1930s, but in Australia and South Africa, it only became common in the 1970s. By 2002, the mining industry had accepted open towns or normalisation as the main way to house its workforce (IIED, 2002), and collaborative planning had become the main way to manage these towns. In Australia, this shift to open towns coincided with the rising popularity of fly-in-fly-out arrangements (Storey, 2001). Worker camps (instead of company towns or open towns) became common in Australia and Africa (though not in South Africa) (Rubbers, 2019). Mining economies in this context often bypass local communities.

The aforementioned shifts in policy for housing miners have implications for mine closure. The mining companies' long-term liabilities have been reduced, and the responsibility for dealing with settlements at closure has shifted to democratically elected governments and households. Local governments have to manage both booms and busts, and householders have to cope as best they can. Policies promoting normalisation largely ignore these implications, as they ignore the long-term reality of mine closure.

Globally, today we see various types of settlements near mines: open towns, company towns and worker camps. South African policy promotes normalisation and the development of open towns (Van der Watt & Marais, 2019; Marais, 2023). This attempts to redress the historical exclusion of black people from mining towns through the migrant labour system. The aim is to bring democracy to company towns, ensure the long-term development of these towns, and link the mining and non-mining populations. Formal governance structures and mining communities become responsible for dealing with booms and busts and eventual closure. Normalising and developing open towns may be considered appropriate responses during a mining boom. However, managing decline and closure in open towns is far more difficult. Although mining companies commonly plan the life cycle of a mine, it is difficult for governance structures and mining communities to do the same.

1.4 Mine closure in Africa

Twenty years ago, the World Bank estimated that 25 large African mines would close (World Bank & IFC, 2002). Planning for closure would be essential to avoid the long-term negative consequences of mining. Although the South African mining context is substantially different from elsewhere in Africa, we highlight some of the main attributes of mine closure across the continent. Our focus here is largely on the commercial mining sector, although artisanal mining also has closure concerns (Bryceson et al., 2020).

Since the 1990s, Africa has seen a substantial increase in mining. Many African governments opened their countries to international investors. While the continent still struggles to respond to these investments, governments overlook the reality of eventual mine closure. Consequently, only a small body of literature focuses on mine closures in Africa (outside of South Africa). In the following text, we draw on the following sources from African literature: Ferguson (1999), Obeng-Odooom (2009), Larmer (2010), Kotsadam and Tolonen (2016), Besa et al. (2020), Gough et al. (2019), Knierzinger and Sopelle (2019) and Mfune et al. (2020).

The economic implications of mine closure and decline are usually extensive, with many people losing their mining jobs. Job losses also extend to the large numbers of people employed in businesses indirectly dependent on mining economies, like restaurants and suppliers to the mine. The economic decline has social implications. It can disrupt the social order, threaten safety and security, and increase crime. Loss of household income can result in school dropouts (an intergenerational consequence), as family members find they cannot afford to keep their children in school. In some cases, schools have simply closed, with teachers also losing their jobs. Mine closures can cause an outflow of people. But this is not always the case, and mine closure can lead to the formation of more households, as can be seen, for example, in Ghana.

Although open towns have become the norm, there are often infrastructure dependencies between mines and towns. A common example is the services commonly provided by local government infrastructure installed and maintained by mining companies. In other examples, mines manage and maintain recreational facilities or support local schools by paying the teachers' salaries. Closure brings an end to this, to the detriment of young people.

Land rehabilitation is an economic necessity. Effective land rehabilitation lays the foundation for new economic opportunities. But many mining areas in Africa are stuck with unrehabilitated land because their governments cannot enforce rehabilitation. Unrehabilitated mines attract small-scale miners, and sometimes this contributes to organised crime.

The effects of the large mining investments of the past 30 years will soon become evident when closure sets in. The environmental and social consequences will be huge, and African governments will typically struggle to enforce legislation for land rehabilitation. Mining localities will have to deal with the social and economic implications with little support from the mining companies or their national governments.

1.5 Mine closure in South Africa

Mining has been the historical backbone of development in South Africa. But increasingly, mines are closing. Figure 1.1 shows mines that have closed or are in care and maintenance; these do not include the 6,000 abandoned mines or closures that the government does not fully recognise. Many towns and cities developed because of mining, with Johannesburg being a prime example. Johannesburg

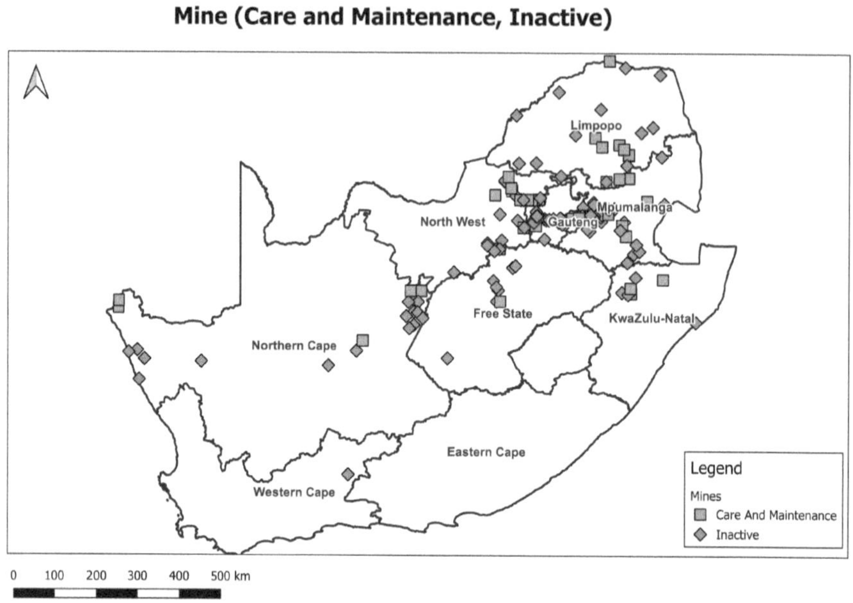

Figure 1.1 Inactive mines and mines under care and maintenance in South Africa

grew from gold, but today it has a diverse economic base, dominated by the service sector. But successful diversification of South African mining town economies is rare. Newcastle has had some success in this endeavour (Marais et al., 2021), but most have struggled. Sol Plaatje (formerly Kimberley, a diamond-mining town), Matjhabeng (Welkom, gold) and Matlosana (Klerksdorp, gold) are examples of towns whose economies have been damaged by mine closure. Other commodities besides diamonds and gold may follow this downward spiral. Towns dependent on coal and platinum are under threat. South Africa's global commitment to cleaner energy sources will affect coal mining areas. The government has already agreed to close most coal-fired stations by 2030 (although they are backtracking on this). Platinum prices are high, but the transition to electric vehicles could reduce demand.

But despite commodity volatility and the looming reality of mine closure, many mining communities are trapped in the expectation of wealth. The temporariness of mining as an economic activity is poorly understood, and the expectation of wealth often continues well after mine closure.

The closure of mine hostels since the early 1990s, when mine housing policy shifted from hostel to family accommodation, has resulted in many people coming to live with the miners near the mines. This policy shift transferred the risks from mining companies to local governments and households. A further policy shift, the drive to give miners homeownership, could be disastrous for the homeowners when decline sets in. Non-homeowning miners' families are free to migrate. They are not dependent on the town prospering. This laudable but perhaps misguided drive is an example of the government's goal dependency, the goal being reparation for apartheid (see Chapter 2).

1.6 The book's motivation, aims and methods

We see mine closure as a global problem that requires context-specific research and a holistic understanding. We had six specific motives for writing this book.

First, we believe the mine community research has focused too exclusively on the relationship between mines and communities during the operational stage of mining. It needs to pay more attention to this relationship at closure and recognise the dependencies that are set up during the operational stage.

Second, mine closure needs to be understood simply as an extension of mine development. Mining companies plan a mine's whole life cycle. But social impact assessments typically focus on mine development, ignoring mining decline. Extending the initial impact assessment to closure could help planners and policy-makers think about closure from the outset.

Third, and connected to the aforementioned point, planners and policymakers must consider mining booms and busts as one process. Treating them as unconnected events means that during a boom, no one considers the possibility of a bust or recognises the dependencies a boom creates. Those dependencies are difficult to deal with when closure sets in.

Fourth, the knowledge base about mine closure is often the property of mining companies and is sectoral and place-specific. The literature needs to be extended beyond the company doors, reflecting on various sectors, places and the responses of communities and governance structures.

Fifth, mine closure research tends to pay more attention to environmental than social aspects of closure. Although both are important and should be seen as interlinked, many policies and guidelines assume that the environmental aspects encompass the social aspects, but often they do not.

Finally, there is a need to collate urban experiences from the South African periphery. We need geographical evidence from remote areas in South Africa to support our research on the social consequences of mine closure in these areas.

The book has four overall aims:

- To describe the nature and scale of mine closure consequences for towns, cities and miners;
- To describe the responses of communities and their organisations to mine closure;
- To analyse existing policy frameworks and responses to mine closure by focusing on social aspects;
- To provide a theoretical framework against which to understand the social aspects of mine closure and decline.

Our methods were both theoretical and practical. The theoretical method is important, as to date the mine closure literature has tended to be atheoretical. We drew on two areas of theory: social disruption and evolutionary governance theory (see Chapter 2 for a detailed account). The concept of 'social disruption' is commonly used to describe the effects of mine growth on communities. We used this concept to investigate the effects of closure. Large mining investments contribute to crime, alcohol and drug abuse, and other social ills. O'Connor and Ruddell (2021), in a case study from North America, found no evidence that closure results in social disruption. Our case studies found evidence that it does, and the concept of 'social disruption' helped us to show that such disruption is a common syndrome, not just isolated problems.

We argue that social problems caused by mine closure are a direct result of dependencies created during the life of a mine. The concept of dependencies is drawn from evolutionary governance theory. This theory helped us to identify common syndromes illustrated by the case studies: long, slow change and inability to change (path dependency), policies created for mining growth that are unsuitable for managing closure (goal dependency), and relationships set up during a boom that are disrupted during a bust (interdependency).

Our practical method was to conduct case studies in a wide spread of mining towns in South Africa, particularly in peripheral areas. Except for the West-Rand study, all are from remote areas across the country. This includes a few secondary cities, small towns, and the rural communities of Tshikondeni and the Richtersveld. Most of the chapters are based on a single case study. We simply required the researchers who contributed the chapters to relate the story of mine closure or decline (or expected decline) in their case study area, to record any social disruption it may have caused, and to consider how the various dependencies have shaped the response to decline. For the concluding chapter, we adopted a comparative approach to analyse the differences and similarities across the case studies.

The various chapters provide brief overviews of specific methods the researchers used. Not all the case studies are of actual mine closure or substantial decline; some are studies of areas with a high likelihood of decline. Figure 1.2 shows the

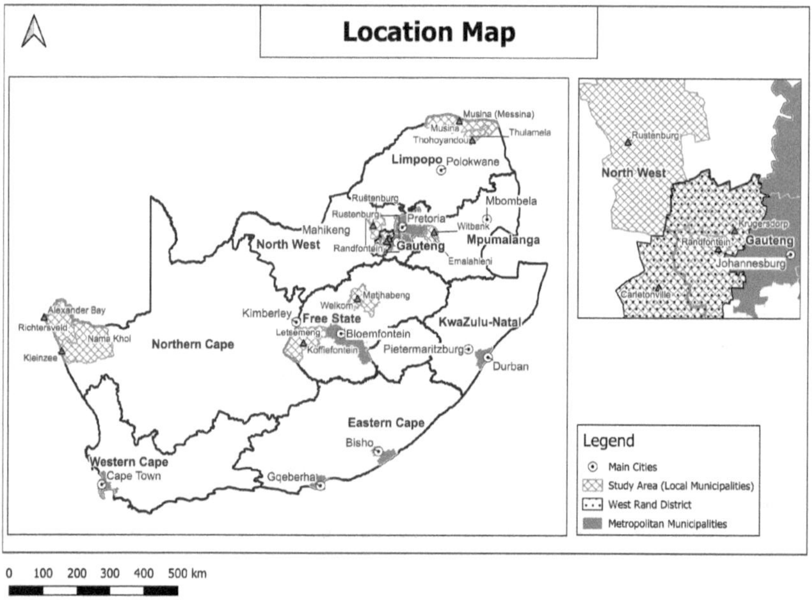

Figure 1.2 Location of case study areas

location of the case studies. In addition to the place-based case studies, we also commissioned a case study of miners' responses to retrenchment (Chapter 5).

1.7 Outline of the book

1.7.1 Section 1: Background and theory

Chapters 2 to 4 provide a theoretical framework, an account of historical and present mine closure legislation in South Africa, and a review of the literature. This general section is followed by the case studies. The first case study looks at the situation of miners retrenched because of closure. The rest are place-specific studies of decline and closure, looking at the consequences and the local responses (or lack of responses) and relating the findings to the book's theoretical framework. The final chapter combines the lessons learned and presents an expanded framework for understanding mine decline and closure.

Chapter 2: A theoretical framework for understanding the social aspects of mine closure

This chapter explains the theoretical framework and discusses its use for explaining the social aspects of mine closure. It assesses current understandings of the social consequences of mine closure against the concept of social disruption and the dependencies identified by evolutionary governance theory. Despite little evidence from research in the Global North to link mine closure and social disruption, we argue that social disruption does occur when mines close. We argue that social disruption occurs because mining creates dependencies that are difficult to overcome when mines close.

Chapter 3: Mine closure policies and strategies in South Africa: a critical review

As elsewhere in the world, mine closure policy in South Africa emphasises environmental aspects. Often, the social aspects are embedded in environmental policy and legislation. The original government policies regarding mining make scant reference to closure. Consequently, South Africa has over 6,000 abandoned mines with little likelihood of environmental restoration or repurposed land use. The chapter traces the development of mine closure legislation and evaluates the 2020 Draft Mine Closure Strategy. Historical legislation broadly referred to restoring land to its original state, but this was never enforced. The Minerals Act (1991) was the first to require an environmental management programme, a rehabilitation plan and financial provision for closure. Since democracy in 1994, several environmental and mining laws have followed, emphasising sustainability, appropriate closure management and post-mining requirements. The 2020 Draft Mine Closure Strategy provides guidelines for mine closure. Although this strategy has some valuable aspects, it still does not pay enough attention to the social aspects.

Chapter 4: A scoping review of the literature on mine closure

This chapter is a literature review that investigates the social implications of mine closure. Many mine closure frameworks have been developed only since the 1990s, but practical implementation and a thorough understanding of the social aspects are lacking. Debates about mine closure often prioritise economic and environmental rather than social concerns. Industry viewpoints dominate the literature, while socio-economic planning tends to be treated as a technical exercise. This chapter reviews some literature on the evolution of mine closure concepts and the emergence of social costs as a critical concern. There is reference to climate implications and their interconnectedness with social dimensions like health implications and cultural alienation.

1.7.2 Section 2: Case studies

Chapter 5: Miners' lives after retrenchment

When a mining company reduces its workforce through collective layoffs or decides to close an entire operation, it can bring hardship to many. The consequences may be worse in developing countries, where the mining sector is an important contributor to economic stability and employment is less secure than in developed countries. The chapter discusses the findings of two studies conducted in South African mining communities. It illustrates the effects of mine downscaling and closure on the lives of miners and their families. It describes the experiences and perceptions of those who lost their jobs and income. It shares their narratives about how they dealt with their loss and what helped them cope. The chapter suggests what could have been done differently in each case. The participants' harrowing accounts of social disruption at the household level indicate how dependent miners are on mine employment.

Chapter 6: West Rand: decline in South Africa's economic heartland

The scale of mine decline on the West Rand over the past three decades has been massive. Historically, the area's economy depended heavily on gold and uranium mining. The West Rand attracted large numbers of migrant labourers, mostly unskilled and low-skilled, from other parts of South Africa and the SADC countries. The mine closure caused large-scale job losses and outmigration to other areas of Gauteng. Many who remained were left with no alternative livelihood. The local economy has been severely affected, and the informal sector has grown rapidly. Informal settlements have burgeoned close to the mines. Mine decline has damaged the environment and human health and contributed to land invasions, social unrest and increased crime levels. Municipal finances have dwindled as the tax base has shrunk and investments in the local economy have declined, making it difficult for municipalities to deliver services. The West Rand struggles to respond adequately to mine decline and to devise appropriate strategies to mitigate its difficulties.

Chapter 7: Matjhabeng: decline in the urban periphery

In the mid-1980s, the Free State Goldfields, now known as Matjhabeng Local Municipality, produced 25% of the gold in the free world and employed 180,000 miners. The decline began in the early 1990s, and today the Goldfields employ only about 20,000 miners. Research has documented this decline's social and economic consequences and the various attempts to revitalise the local economy. In 2019, the municipality released yet another of its many plans to diversify the economy. The chapter assesses this latest plan against the international literature. The new plan prioritises plan-making over implementation, underplays the need for local actors to engage in good governance, ignores existing equilibriums and fails to consider the area's regional services function as a form of economic diversification.

Chapter 8: Kleinzee: looking for a new dawn amidst the diamond dust

Kleinzee is a small coastal town in the Northern Cape Province. Mining of alluvial diamonds started here in the mid-1920s. The mining company De Beers was instrumental in the town's founding and the region's day-to-day operations. At one point, De Beers owned about 320 km of coastal land south of Port Nolloth. Official diamond mining activity declined in the mid-1990s, and De Beers stopped operating in 2012. Today, large portions of land are still owned by De Beers, while local communities and the region struggle to deal with the economic and social consequences of mine closure. The chapter considers the socioeconomic impact of the official closing of the De Beers diamond mine. It looks at how the 400 residents remaining from the original 4,000 are experiencing this impact amidst frequent reports of illegal mining. It looks at the perspectives, experiences and aspirations of the youth in a town that once hosted three primary schools and an Olympic-size swimming pool, now empty. The chapter looks for answers in official publications and personal interviews with relevant stakeholders.

Chapter 9: Koffiefontein mine downscaling: socioeconomic and infrastructural consequences

The chapter looks at the livelihood choices and practices of Koffiefontein residents, ex-miners and business owners and compares their current diverse range of livelihood activities with the outcomes of a social impact assessment conducted in 2005. It describes the infrastructure of municipal services such as water, sewers, roads, stormwater drainage and electricity, which need ongoing maintenance, upgrading and replacement. With the series of mine downscalings over the years, the infrastructure has become obsolete or requires increased maintenance expenditure. Mitigating the socioeconomic and infrastructural consequences of mine downscaling in Koffiefontein requires creative solutions, community collaboration and appropriate technology.

Chapter 10: Alexkor and the Richtersveld community: unlikely partners to mine diamonds together

In 2003, the Richtersveld community successfully obtained the restitution of diamond-rich land along the northern tip of South Africa's West Coast. This resulted in a Pooling and Sharing Joint Venture (PSJV) with the state mining company Alexkor. The Richtersvelders got a 49% share in an almost bankrupt mine company. The internal conflict caused the governance system to collapse, leaving the Richtersvelders without representation in the PSJV. Alexkor thus practically runs the PSJV. Alexkor has been hampered by national agendas, mismanagement, corruption, loss-making and state capture. By 2020, both partners were under administration, and the PSJV was bankrupt. Interviews with community members point to social disruption, conflict and mistrust. Local contractors lost their investments through high-risk contracts with the PSJV. The actors are now locked into interwoven forms of dependency: a court order binding them into a PSJV that has been dysfunctional since inception, the state's ideological agendas, and decades of mistrust and interdependence between Alexkor and the Richtersvelders. The community is trapped in internal conflict, yearning for historical justice and dreaming of immense wealth from their diamond-rich land.

Chapter 11: Tshikondeni: mine closure in a deeply rural area

The chapter investigates the effects of the 2014 closure of the Tshikondeni coal mine in Musina Municipality in the far north of Limpopo province. The authors interviewed community members and traditional leaders. The Exxaro mining company declined to be interviewed, so information was gleaned from its website. The findings reveal three different interpretations of the effect of the mine closure. Exxaro said the closure had followed all the prescribed regulations, including reskilling and offering positions at other mines. The traditional leaders took a far more negative view of the social and economic consequences of the mine closure for their community. The community had mixed views but was concerned about the socioeconomic consequences of declining jobs, income and services. While the mine rehabilitates the environment, the villagers are damaging it by cutting down trees to sell for firewood, as they have no alternative livelihoods.

Chapter 12: Emalahleni's just transition: from closure to collaboration?

Having signed the Paris Agreement, South Africa is under pressure to change from fossil fuels to renewable energy sources. Emalahleni, in the Mpumalanga province of South Africa, is home to 80% of South Africa's coal reserves, 15 coal mines, and six coal-fired power plants, with about 35,000 people formally employed in the sector. The challenge is how to make the change to renewable energy sources a success for the people of Emalahleni. Any transition that benefits the environment but not the people is doomed to failure. The chapter describes the social impacts of closing Emalahleni's coal-fired plants and a possible decline in some coal mines. A post-coal future will have to consider the dependencies that have been caused by the dominance of the coal mining sector. The question is how to aim for local stewardship, community care and caring for the environment in such a contested context. The chapter considers the difficulties and the realistic possibilities for a just transition in Emalahleni.

Chapter 13: Rustenburg: the struggle to go beyond mining

The platinum mining industry grew rapidly in South Africa for about three decades up to 2011. After the repeated labour strikes of 2012 and 2014, production dropped significantly, and the primarily mine-related local economy suffered. As a result, the area has been experiencing business closures and high unemployment. Over the past five years, the platinum price has buffered some closure problems. Using key informant interviews and document analysis, the chapter shows how the Royal Bafokeng Nation has profited from mining and used the profits to benefit the community, particularly by promoting education. It shows how the mining companies have left the municipality to deal with the consequences of closure. Despite formulating plans for a 'Rustenburg beyond mining', the municipality's overall approach to economic diversification has been merely reactive. The chapter foregrounds the notion that dependencies created by mining can be turned into opportunities when a multi-stakeholder, participatory approach to mine downscaling is adopted.

Chapter 14: Rustenburg: coping despite mine decline

Since 1979, Rustenburg's economy has been driven by the mining of platinum group metals, but since 2007, despite some booms, it has gone into decline. The chapter is based on a case study of Freedom Park, a low-income area of Rustenburg Local Municipality. The ordinary people of this area have shown resilience and resourcefulness despite the challenges of mining volatility and economic restructuring. They have given their children access to higher education. Their efforts have intergenerational value: their children's qualifications will enable them to find better jobs than their parents had, thus helping to diversify household livelihoods and minimise the long-term social consequences of mine closure. The chapter reflects on what further decline will mean for Rustenburg, but the narratives from the interviews indicate that ordinary people will cope.

Chapter 15: A framework for understanding the social aspects of mine closure in South Africa

In this chapter, the editors return to the theoretical concepts used in the book. We summarise examples of these concepts we have identified from the case studies, assess the scale and nature of social disruption, and link this to the dependencies associated with mining and mine closure. The chapter broadens the discussion on the theoretical aspects. We present our extended version of the framework in Chapter 1, giving a more comprehensive overview of the social aspects of mine closure. We highlight the consequences of mine closure and the responses from communities, miners and local organisations. We suggest alternative ways to understand the social cost of mine closure and make policy recommendations.

References

Bainton, N. & Holcombe, S., 2018. A critical review of the social aspects of mine closure. *Resources Policy*, 59, pp. 368–478.

Besa, B., Kabwe, J.M.J. & Banda, W., 2020. Socio-economic impact of mine closure and development of exit strategy for rural mining areas in Zambia: A case study of Kalumbila District. In *Proceedings of the 28th International Symposium on Mine Planning and Equipment Selection – MPES 2019*. Berlin/Heidelberg: Springer.

Bryceson, D., Jønsson, J. & Shand, M., 2020. Mining mobility and settlement during an East African gold boom. *Mobilities*, 15(3), pp. 446–463.

Chaloping-March, M., 2017. *Social Terrains of Mine Closure in the Philippines*. London: Routledge.

Ferguson, J., 1999. *Expectations of Modernity: Myths and Meanings of Urban Life on the Zambian Copperbelt*. Berkeley, CA: University of California Press.

Gough, K., Yankson, P. & Esson, J., 2019. Migration, housing and attachment in urban gold mining settlements. *Urban Studies*, 56(13), pp. 2670–2687.

IIED (International Institute for Environment and Development), 2002. *Breaking New Ground: Mining, Minerals and Sustainable Development (MMSD)*. London: Earthscan.

Knierzinger, J. & Sopelle, I., 2019. Mine closure from below: Transformative movements in two shrinking West African mining towns. *Extractive Industries and Society*, 6, pp. 145–154.

Kotsadam, A. & Tolonen, A., 2016. African mining, gender, and local employment. *World Development*, 83, pp. 325–339.

Larmer, M., 2010. Historical perspectives on Zambia's mining booms and busts. In Fraser A. & Larmer, M. (eds.), *Zambia, Mining and Neoliberalism: Boom and Bust on the Globalised Copperbelt*. New York: Palgrave Macmillan, pp. 31–58.

Marais, L., 2023. *The Social Impacts of Mine Closure in South Africa: Housing Policy and Place Attachment*. London: Routledge.

Marais, L., McKenzie, F., Deacon, L., Nel, E., Van Rooyen, D. & Cloete, J., 2018. The changing nature of mining towns: Reflections from Australia, Canada and South Africa. *Land Use Policy*, 76, pp. 779–788.

Marais, L., Nel, V., Rani, K., Van Rooyen, D., Sesele, K., Van der Watt, P. & Du Plessis, L., 2021. Economic transitions in South Africa's secondary cities: Governing mine closures. *Politics and Governance*, 9(3), pp. 381–392.

Mfune, O.K.-W.C., Chansa-Kabali, T., Chisola, C. & Manchisi, J., 2020. The legacy of mine closure in Kabwe, Zambia: What can resilience thinking offer to the mining sustainability discourse? In Gasparatos, A., Naidoo, M., Ahmed, A., Karanja, A. & Fukushi, K. (eds.), *Sustainability Challenges in Sub-Saharan Africa: Insights from Eastern and South Africa*. Singapore: Springer Nature, pp. 167–188.

Ndaguba, E. & Marais, L., 2023. A scientometric analysis of mine closure research. Environment, Development and Sustainability. (Accepted).

Neil, C., Tykkylainen, M. & Bradbury, J. (eds.), 1992. *Coping with Closure: An International Comparison of Mine Town Experiences*. London: Routledge.

Obeng-Odoom, F., 2009. Oil and urban development in Ghana. *African Review of Economics and Finance*, 1(1), pp. 18–39.

O'Connor, L. & Ruddell, R., 2021. After the downturn: Perceptions of crime and policing in the southeastern Saskatchewan oil patch. *The Canadian Geographer*, 5(3), pp. 281–291.

Rubbers, B., 2019. Mining towns, enclaves and spaces: A genealogy of worker camps in the Congolese copperbelt. *Geoforum*, 98, pp. 88–96.

Storey, K., 2001. Fly-in/fly-out and fly-over: Mining and regional development in Western Australia. *Australian Geographer*, 32(2), pp. 133–148.

Van Assche, K., Gruezmacher, M. & Granzow, M., 2021. From trauma to fantasy and policy: The past in the futures of mining communities; the case of Crowsnest Pass, Alberta. *Resources Policy*, 72, p. 102050.

Van Assche, K., Gruezmacher, M., Marais, L. & Perez Sindin-Lopez, Z., 2024 (in press). *Resource Communities: Past Legacies and Future Pathways*. London: Routledge.

Van der Watt, P. & Marais, L., 2019. Normalising mining company towns in Emalahleni, South Africa. *The Extractive Industries and Society*, 6, pp. 1205–1214.

Vivoda, V., Kemp, D. & Owen, J., 2019. Regulating the social aspects of mine closure in three Australian states. *Journal of Energy & Natural Resources Law*, 37(4), pp. 405–424.

Watson, A. & Olalde, M., 2019. The state of mine closure in South Africa – What the numbers say. *Journal of the Southern African Institute of Mining and Metallurgy*, 119, pp. 639–645.

World Bank & IFC (International Finance Corporation), 2002. *It's Not Over when It's Over: Mine Closure Around the World*. Washington, DC: World Bank Group's Mining Department, Global Mining.

2 A theoretical framework for understanding the social aspects of mine closure

Lochner Marais, Sethulego Matebesi and Verna Nel

2.1 Introduction

The mine closure literature has some shortcomings. Notably, it has tended to focus too exclusively on environmental concerns. Our particular concern here is that work on the social aspects is in short supply and mostly applied and atheoretical. In this chapter, we suggest two theoretical approaches to fill these deficiencies: social disruption and evolutionary governance theory. Notable among the few recent studies that do consider the social aspects of mine closure are a critical assessment (Bainton & Holcombe, 2018), a study of closure policy (Vivoda et al., 2019) and two single-authored books (Chaloping-March, 2019; Marais, 2023).

Theoretical frameworks against which to understand the social effects of mine closure remain scarce. Digby (2012, p. 1) argues that the concept of deindustrialisation provides such a theoretical framework for understanding how communities can be 'equipped for life after mining'. Yet several peculiarities of mine closure complicate the use of this concept to explain the social effects. For example, unlike deindustrialisation in manufacturing, mine closures are not always the result of global competition. A second, related, concept is the 'shrinking city' (Martinez-Fernandez et al., 2012; He et al., 2016). The literature on shrinking cities emphasises the advantages of decline and downsizing and advocates planning for it. These two concepts are useful, but limited. In a recent book, Marais (2023) adds depth by considering what the social disruption concept and the dependencies that are a central concept in evolutionary governance theory can tell us about the social impacts of mine closure. He criticises the developmental approach to mining and housing in South Africa.

Why should we be concerned about theorising our research? Damage to people and communities by the closing of a mine is surely a practical issue. The answer is that our analyses and arguments need to have a basic set of statements to which researchers can refer to ensure that we are not talking at cross purposes but are in effect 'on the same page', that we have some common reference points. We need common ground so we can identify the principles at work behind the many different examples we see of mine closure. Research needs a theoretical framework so we will be able to say not just *what* happened but *why* it happened and to recommend policy that is principled rather than ad hoc. It is this framework that makes

DOI: 10.324/9781003403326-2

our book more than just a collection of anecdotes. The chapters do not always make a big point of referring to the framework, but it is there behind the authors' thinking.

This chapter discusses two theoretical frameworks: the social disruption concept and the dependencies identified by evolutionary governance theory. We see our two frameworks as connected. We show in this book that dependencies lead to social disruption when a mine declines. But perhaps it was the disruption caused by a mining boom, the loss of traditional routines, that led to the dependencies in the first place. For example, a mining boom in a South African city causes social disruption by bringing large numbers of labourers to the city. In response, policymakers set the goal of homeownership to create stability. But dependence on this goal turns out to be a mistake when the mine declines, as this book will show.

The goal dependencies, path dependencies and interdependencies that resulted from efforts to cope with a mining boom are difficult to undo when the mine declines. These dependencies that were previously enabling become a hindrance when they lose their usefulness during a bust. They cause new social disruption. Like booms and busts, disruptions and dependencies can be seen as cyclical.

2.2 Social disruption

2.2.1 Background and definition

'Social disruption' is not a formalised theory but a concept used in sociology. In the field of research into the social effects of mining, the term has usually been applied to the disruptive effects of a mining boom. For England and Albrecht (1984, p. 234), social disruption is a reaction to 'a period of generalised crises and loss of traditional routines and attitudes' associated with rapid growth (often mining-related). It is often loosely used to describe a change or breakdown in the social life of a community caused by a 'shock' of some kind, a large and sudden happening such as economic growth or decline, a mining boom or bust, a natural disaster, demographic change, population increase or decrease, a rapidly adopted technological development, a change in government policy or an outbreak of violent crime. England and Albrecht's definition connects social disruption with rapid growth. We show in this book that it is also connected with rapid decline.

Shaw and McKay (1942) first used the concept of social disruption to explain youth delinquency in Chicago. They used the term 'social disorganisation' to argue that bad neighbourhoods create delinquent youth behaviour. Although 'social disruption' is commonly used as a sociological term today, it originates from psychological or individual understandings of disruption. Goffman (1971) made the first attempt to translate the concept of individual disruption into a sociological concept. Disruption is caused by individuals who do not behave according to the social understandings, or codes of conduct, that are necessary for society to function in an orderly fashion. A sociological understanding emphasises that disruption emerges from social situations or external factors. Volimer (2013, p. 7) argues that there is room for improved theorisation about disruption from a sociological perspective and a need for 'systematically accumulating conceptual and empirical intelligence about disruptions'.

Although disruption often occurs in the normal course of events, it gained global traction in the mining context in the 1970s and 1980s, when large mining and energy projects disrupted small urban centres or rural areas.

2.2.2 Social disruption in mining town research

The concept of social disruption has been central to social impact assessment (Finsterbusch & Freudenburg, 2002), and Jones and Mayzer (2021) use it as a central theme to analyse boomtown research in the mining context.

Jones and Mayzer (2021) refer to five phases of boomtown research, four of which were originally identified by Finsterbusch and Freudenburg (2002). In Phase 1 (the 1960s), a mining boom was considered a positive phenomenon. It created local jobs and ensured business and housing booms. Impact assessments ignored the adverse effects. By the 1970s (Phase 2), research showed that mining booms could have negative local implications (Little, 1976). The economic burden resulting from a lack of infrastructure in the usually remote locations was central to understanding the adverse effects. The rapid inflow of people resulted in housing shortages, increased crime, school dropouts, a rising divorce rate, and alcohol and drug abuse.

With Phase 3 in the early 1980s, the pendulum swung again. Wilkinson et al. (1982, p. 275) criticised the research on Phase 2 as 'undocumented assertions, questionable interpretations of evidence, and superficial analyses'. There had been a tendency to overstate the negative social effects, and there were methodological concerns. Questions were asked about the relevance of the indicators, small sample sizes and lack of longitudinal studies. Phase 3 researchers promoted a more balanced approach. They noted that boomtown experiences could vary across different groups, and they challenged Phase 2's negative perspective on local economic growth.

Phase 4, at the end of the 1980s, took a far more balanced approach. Researchers assessed both positive and negative effects and highlighted their complexities. More longitudinal research was done in response to the methodological criticism. Researchers accept that the scale and nature of social disruption vary and that different groups have different experiences (England & Albrecht, 1984).

Jones and Mayzer (2021) initiated a fifth phase, emphasising the complexities associated with social disruption. This phase takes a more critical perspective. It adds 'thematic explorations' and 'the public health perspective', looks at 'how gender intersects with issues unique to the boomtown setting' and views boomtowns 'through the lens of late modernity' (Jones & Mayzer, 2021, p. 232). New themes that emerged included gender, the complexities associated with housing, Indigenous communities and the shift towards individuals and households for dealing with the negative implications.

The first four phases of the boomtown research were done mainly in the United States. One longitudinal study from Australia (Smith et al., 2001) found that social disruption declines over time. In Phase 5, a substantial body of work has emerged from Australia and Canada.

2.2.3 Social disruption and decline

In the Global North, the social disruption concept has been used to explain the social consequences of a mining boom but not of a mining bust. In contrast, we have found that it explains the effects of mine closure in the Global South (Marais et al., 2022; Sesele & Marais, 2022).

The literature on mining booms is large. Far less has been written on decline. Mining busts have not been associated with social disruption. O'Connor and Ruddell (2021), for example, in a survey conducted in Canada, found no evidence that crime had increased in areas where oil mining had declined. The paucity of evidence from the Global North may be because social security programmes mitigate the adverse effects of mine closure. It may also be because the more diverse economic profiles of countries and technologically advanced ways of mining in the Global North help people find work in other sectors. There is, however, contrasting evidence from the Global South pointing to social disruption from mine closure. Research from South Africa shows that mining areas in decline have higher crime levels than fast-growing mining areas (Marais et al., 2022; Sesele & Marais, 2022; Marais, 2023). These studies of social disruption associated with decline could launch a sixth phase of boomtown research. This phase will show that booms and busts are two sides of the same coin and that much of the disruption during a decline results from the dependencies created during a boom.

2.3 Evolutionary governance theory and the dependencies

2.3.1 Background and definition

The concept of dependencies is not new but has been systematised by evolutionary governance theory. This theory of governance emphasises that change is slow (Van Assche et al., 2014). The definition of 'governance' that we use in this book is 'process by which governments (local, regional and national) and stakeholders collectively decide how to plan, finance and manage urban areas' (Avis, 2016, p. 1). 'Governance' has a broader meaning than 'government'. It includes the roles of the private sector and community organisations.

The foundations of evolutionary governance theory lie in new institutional economics, complexity theory, biological evolutionary theory, poststructuralism, social systems theory and actor-network theory. The theory focuses on non-linear causality, promotes a middle ground for political-economic ideologies and emphasises slow change (evolution). It has been used to help explain the governance of mining and post-mining societies (Marais et al., 2020; Van Assche et al., 2017). Generally, mining booms and busts happen rapidly, while governance responses are slow. Evolutionary governance theory says that the slowness is because policymakers are held up by path dependencies, goal dependencies and interdependencies. We explain these three dependencies in the following and then look at their application to mining towns and at strategic planning, or 'steering' as an attempt to overcome these dependencies.

2.3.2 The dependencies

Evolutionary governance theory identifies three types of dependency: goal dependency, path dependency and interdependency. The setting of objectives by organisations to create change is a common management process. But the literature has taken little notice of the way such objectives can become stumbling blocks. Van Assche et al. (2014) attribute this to goal dependency. The United Nations' 17 Sustainable Development Goals are a shared vision for the world. Yet some of these goals may become irrelevant or less important in specific locations, and depending on them to solve problems may be a mistake. For example, indicators to upgrade slums can easily lead to relocation (an example of a goal dependency). Marais (2023) argues that post-apartheid policies that undo migrant labour and emphasise homeownership are examples of goal dependency; their goal is to undo the ills of apartheid. He shows how this goal, though laudable, can have adverse consequences for miners and households in declining mining towns. For example, the laudable goal of breaking the apartheid migrant labour system and encouraging homeownership becomes a crippling dependency for labourers when the mine closes and they are left with an unsellable home. Bryceson and MacKinnon (2013) say that policy approaches for these towns should be flexible and acknowledge the temporariness of mining.

The setting of goals has been closely associated with the concept of steering. Steering refers to the ability of the government and partners to respond to realities through policy, strategy and planning (Beunen & Van Assche, 2021). This would imply ensuring that the government's chosen policy goals are appropriate. To govern and plan for booms and busts requires flexible strategies. Despite criticism of the ability of steering to bring about changes, there is an 'ongoing quest for policies to steer and adapt' (Beunen & Van Assche, 2021, p. 365). Goal dependency, not necessarily a bad thing, can become goal fixation, blinding policymakers to other opportunities or the negative consequences of the goal.

A related concept is path dependency, which results from dependence on a goal. Evolutionary governance theory borrowed the term from historical institutionalism (Krasner, 1984) and new institutional economics (North, 1990, 2005). It is used to explain how historical decisions influence current ones. This historical dependence is one of the reasons change is often slow. Path dependency can contribute to the inability to change or lock-ins (North, 2005). The persistence of the QWERTY keyboard, although it may no longer be the most efficient design, is a familiar example of a path dependency. A mining boom can have the same effect by creating the impression of a path of long-term prosperity. The result is that policymakers and local communities either ignore a bust or are taken by surprise. Once decline sets in, it becomes difficult to manage. Policies based on growth continue, on the assumption that the path of growth will continue after the mine closes. Path dependency explains why people continue doing things the way they have always been done and cannot find alternatives.

The relationships between actors sometimes lead to interdependency, restricting independent action. To analyse the case histories in this book, we have selected only the concept of dependencies. Van Assche et al. (2022) note that 'Dependencies are also enabling and potentially, even guiding factors'. But in this book, we look only at negative dependencies. We see in our case histories the damaging effect of dependencies that have lost their usefulness. Interdependencies can be hostile or cooperative. Economic activity requires cooperation. For example, publishing a book requires cooperation between an author, a publisher, a printer, contract attorneys and others. The parties are interdependent because of their shared goal: producing the book. Evolutionary governance theory is complex. But interdependency can also result from competition, conflict or a lack of trust. The case histories in this book provide ample evidence of a lack of trust between mining companies and communities (Marais et al., 2021). Mine closure processes exacerbate this lack of trust, and relationships become even more complicated. Interdependency, the mutual dependence between two or more entities, can thus be positive or negative. In the mining environment, mine closure often means that the mining company breaks the interdependence and withdraws, resulting in social disruption.

Evolutionary governance theory has implications for boomtown research and for our topic in this book: mining decline (Van Assche et al., 2014, 2017). Slow policy change because of dependencies means that responses to local situations could result in governance steering having unintended consequences. Overcoming these dependencies, caused originally by a boom, is essential for finding alternative responses to mine closure.

2.4 Conclusion

The chapter provided an overview of the two conceptual frameworks for this book: social disruption and evolutionary governance theory. We showed how these frameworks are linked, and we suggested how they might be useful for explaining the social consequences of mine closure.

The case histories provide ample material for testing this usefulness. They provide evidence of social disruption caused by mining decline in the Global South, contrary to the lack of evidence of such a consequence in the Global North and to the general assumption that a boom rather than a bust is the cause of social disruption. They thus extend the social disruption concept. We will also show in the concluding chapter that many of the unfortunate consequences of mine closure that the case histories describe can be explained in terms of the three dependencies, thus providing a solid theoretical basis for policy recommendations. We highlight the following main implications.

First, there is enough evidence to counter research from the Global North showing little social disruption during mine closure. Mining busts in the Global South are different. For example, most countries in the Global South do not have social support systems. Furthermore, mining in the Global North is more technologically advanced. This technological advancement improves finding alternative employment, as some technical skills might be transferable to other sectors. Finally, mining is usually a dominant economic sector in many countries in the Global South. The ripple effect of closure is more problematic in countries with diversified economies.

Second, the chapter links decline and closure to booms. Booms and busts are a continuum and not simply two sides of the same coin, and mining booms create dependencies. For example, understanding the reasons for booms might help plan declines. Moreover, if we manage booms better, the decline might be easier or less disruptive. Mining booms might be why mining decline is difficult or the reason behind social disruption at closure. For example, South Africa's government deliberately links mining with local development (Marais, 2023). This link assumes long-term economic prosperity and does not consider the likelihood of closure.

Another example is the emphasis on homeownership in responses to mining instability. Housing and settlement development have to mitigate the disruption of mining booms. Yet this approach is the main reason decline will create social disruption. It is a good example of a goal dependency created during booms. A further advantage of using evolutionary governance theory is considering how different organisations respond differently over time to booms and busts. For example, how decisions during booms are counterproductive when busts set it. But it also helps us to understand busts and the decisions taken during busts.

Finally, we try to link mining dependencies and social disruption in this book. The dependencies created by mining often continue during decline and closure. These dependencies are central to understanding the social disruption of mine decline and closure.

References

Avis, W., 2016. *Urban Governance.* (Topic guide). Birmingham: GSDRC (Governance and Social Development Resource Centre), International Development Department, College of Social Sciences, University of Birmingham.

Bainton, N. & Holcombe, S., 2018. A critical review of the social aspects of mine closure. *Resources Policy*, 59(C), pp. 368–478.

Beunen, R. & Van Assche, K., 2021. Steering in governance: Evolutionary perspectives. *Politics and Governance*, 9(2), pp. 365–368.33

Bryceson, D. & MacKinnon, D., 2013. Eureka and beyond: Mining's impact on African urbanisation. *Journal of Contemporary African Studies*, 30(4), pp. 513–537.

Chaloping-March, M., 2019. Mine closure: Anticipating the end before the beginning. In Chaloping-March, M. (ed.), *Social Terrains of Mine Closure in the Philippines*. Abingdon: Routledge, pp. 1–16.

Digby, C., 2012. Mine closure through the 21st century looking glass. In Fourie, A.B. & Tibbett M. (eds.), *Mine Closure 2012: Proceedings of the Seventh International Conference on Mine Closure*. Perth: Australian Centre for Geomechanics, pp. 33–38. https://doi.org/10.36487/ACG_rep/1208_04_Digby

England, J. & Albrecht, S., 1984. Boomtowns and social disruption. *Rural Sociology*, 49(2), pp. 230–246.

Finsterbusch, K. & Freudenburg, W., 2002. Social impact assessment and technology assessment. In Dunlap, R. & Michelson, W. (eds.), *Handbook of Environmental Sociology*. Westport: Greenwood Press, pp. 287–299.

Goffman, E., 1971. *Relations in Public: Microstudies of the Public Order*. Harmondsworth: Penguin.

He, S., Lee, J., Zhou, T. & Wu, D., 2016. Shrinking cities and resource-based economy: The economic restructuring in China's mining cities. *Cities*, 70, pp. 75–83.

Jones, M. & Mayzer, R., 2021. A fifth wave? The changing nature of boomtown research. *The Extractive Industries and Society*, 8(2), p. 100893. https://doi.org/10.1016/j.exis.2021.02.013

Krasner, S., 1984. Approaches to the state: Alternative conceptions and historical dynamics. *Comparative Politics*, 16(2), pp. 223–246.

Little, R., 1976. Some social consequences of boom towns. *North Dakota Law Review*, 53(4), pp. 401–425.

Marais, L., 2023. *The Social Impacts of Mine Closure in South Africa: Housing Policy and Place Attachment*. London: Routledge.

Marais, L., Campbell, M., Denoon-Stevens, S. & Van Rooyen, D., 2021. *Mining and Community in the South African Platinum Belt: A Decade after Marikana*. New York: Nova Science.

Marais, L., Denoon-Stevens, S. & Cloete, J., 2020. Mining and urban sprawl in South Africa. *Land Use Policy*, 98, p. 103953.

Marais, L., Ndaguba, E., Mmbadi, E., Cloete, J. & Lenka, M., 2022. Mine closure, social disruption and crime in South Africa. *The Geographical Journal*, 188(3), pp. 383–400.

Martinez-Fernandez, C., Chung-Tong, S., Taira, N. & Vargas-Hernández, J., 2012. The shrinking mining city: Urban dynamics and contested territory. *International Journal of Urban and Regional Research*, 36(2), pp. 245–260.

North, D., 1990. *Institutions, Institutional Change and Economic Performance*. Cambridge: Cambridge University Press.

North, D., 2005. *Understanding the Process of Economic Change*. Princeton: Princeton University Press.

O'Connor, C. & Ruddell, R., 2021. After the downturn: Perceptions of crime and policing. *The Canadian Geographer*. https://doi.org/10.1111/cag.12671

Sesele, K. & Marais, L., 2023. Mine closure, women and crime in Matjhabeng, South Africa. *Geographical Research*, 61, pp. 18–31.

Shaw, C. & McKay, H., 1942. *Juvenile Delinquency and Urban Areas*. Chicago: University of Chicago Press.

Smith, D., Krannich, R. & Hunter, L., 2001. Growth, decline, stability, and disruption: A longitudinal analysis of social well-being in four western Australian communities. *Rural Sociology*, 66(3), pp. 425–450.

Van Assche, K., Beunen, R. & Duineveld, M., 2014. *Evolutionary Governance Theory: An Introduction*. Heidelberg: Springer.

Van Assche, K., Deacon, L., Gruezmacher, M., Summers, R., Lavoie, S., Jones, K., Granzow, M., Hallstrom, L. & Parkins, J., 2017. *Boom & Bust: Local Strategies for Big Events: A Community Survival Guide to Turbulent Times*. Groningen/Edmonton: In Planning and University of Alberta, Faculty of Extension, Alberta. https://doi.org/10.17418/B.2017.9789491937330

Van Assche, K., Duineveld, M., Gruezmacher, M., Beunen, R. & Valentinov, V., 2022. How can we govern if we don't see our feet? Speaking of the matter of sustainability transitions. *Journal of Environmental Policy & Planning*, 24(3), pp. 277–280.

Vivoda, V., Kemp, D. & Owen, J., 2019. Regulating the social aspects of mine closure in three Australian states. *Journal of Energy & Natural Resources Law*, 37(4), pp. 405–424.

Volimer, H., 2013. *The Sociology of Disruption, Disaster and Social Change*. Cambridge: Cambridge University Press.

Wilkinson, K., Reynolds, R., Thompson, J. & Ostech, J., 1982. Local social disruption and western energy development. *Pacific Sociological Review*, 25(3), pp. 275–296.

3 Mine closure policies and strategies in South Africa

A critical review

Samkelisiwe Khanyile and Lochner Marais

3.1 Introduction

South Africa has more than 6,000 ownerless or derelict mines. Many of these were developed and inadequately closed during the colonial and apartheid periods when the requirements for mine closure and land rehabilitation were not extensive. Post-apartheid policy legislates mine closure, but mines and the government are still guilty of not closing mines. Companies often delay closure by placing mines in care and maintenance or selling them to smaller firms. The government hinders the process by not providing closure certificates or by moving the goalposts. There are no closure standards. The Draft National Mine Closure Strategy (2021) attempts to speed up mine closure processes (DMRE, 2021). This is a step in the right direction, but several concerns remain.

How governments and practitioners conduct social impact assessments has changed (Freudenburg, 1982; Jones & Mayzer, 2021). In the 1960s and early 1970s, they focused on the positive effects of developments like mining. Then the pendulum swung, and by the early 1980s, attention was on the negative effects. By the mid-1980s, researchers were questioning the overly negative approach. Consequently, in the 1990s and 2000s, a more nuanced approach was adopted that emphasised the complexities associated with gender, Indigenous communities and the nuanced nature of impacts. Social impact assessments are usually conducted before large infrastructure investments, and they seldom consider closure (in mining cases). We think that this is a mistake. These assessments should consider the long-term implications of investments. Marais et al. (2022) argue that mine closure represents a sixth phase of boomtown research and that the assessment of mining impacts should include the consequences of closure.

In this chapter, we analyse the 2021 Draft National Mine Closure Strategy against the background of mine closure policies in South Africa since 1956. We show how, despite integrating social and environmental aspects, the draft strategy makes simplistic assumptions about social issues and developing post-mining economies. It does not consider how the large-scale investments that are made when mining starts (often to ensure local credibility) eventually affect mine closure. A closure strategy that does not account for the dependencies created by mining and the social disruption of mine closure will have limited relevance. We argue

DOI: 10.324/9781003403326-3

that an appropriate closure strategy would be to avoid creating dependencies in the first place.

3.2 The history of mine closure in South Africa

Appropriate mine closure is a crucial part of a mine's lifecycle, as mining is by its nature a temporary activity. At closure, both short- and long-term environmental, economic and social consequences should be minimised. Figure 3.1 lists mining-related legislation and policy documents and shows how governments have tried to regulate mining activities since the 1950s.

Early mining legislation was lax. The first strict regulations came in 1956 when the government promulgated the Mines and Works Act (Union of South Africa, 1956). This act required mining companies to provide employee housing, health-care and education and made the first reference to closure. It introduced strict regulations for mine safety, health and worker welfare but said little about the environment and mine closure. Mining companies were required to comply with only minimal environmental management requirements and basic rehabilitation plans. For example, fencing off the mining areas was sufficient to obtain a closure certificate. This legislation was concerned only with the surface rehabilitation of post-mining land and did not refer to below-surface rehabilitation (stabilising tunnels etc. for the long term) or the social aspects of closure.

The Atmospheric Pollution Prevention Act (Republic of South Africa, 1965) regulated air pollution in the mining industry. This act included a reference to mine closure and rehabilitation. Mine operators had to submit a plan to prevent and monitor atmospheric pollution. The Mining Rights Act (Republic of South Africa, 1967) required mine operators to obtain a mining licence, submit a plan for mine

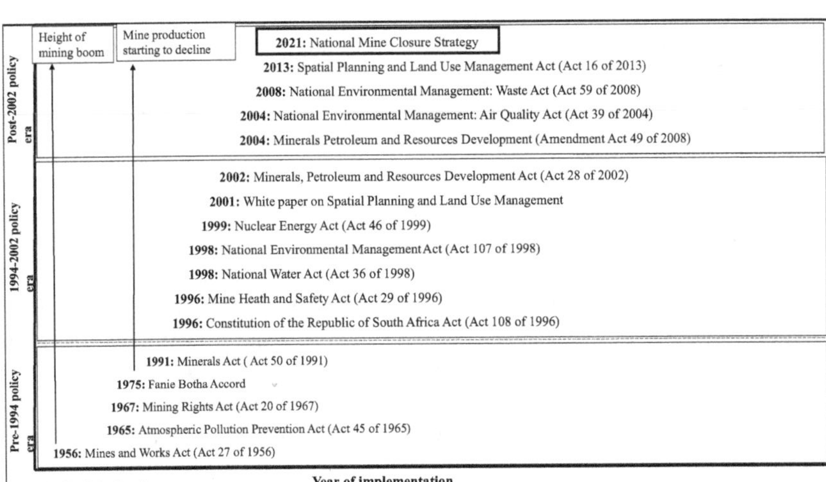

Figure 3.1 Mining-related legislation, 1956–2021

closure and rehabilitation and make financial provisions for environmental rehabilitation, but it did not refer to the social aspects of closure. Despite these requirements, rehabilitation plans were seldom implemented. The Fanie Botha Accord of (Republic of South Africa, 1975) was a further effort to promote responsible mining, including mine closure and rehabilitation (CSIR, 2019). In the Accord, mining companies and the government recognised the environmental impacts and the need for land rehabilitation at closure. The Accord said mines that had closed before 1956 were the state's responsibility. The Accord lacked legally binding provisions, and mining companies continued to operate with minimal legal requirements for the environment or closure.

The Minerals Act (Republic of South Africa, 1991) was the first legislation with closure requirements. Until then, mines could abandon their operations once they became unprofitable. The aim of the act was to ensure that the state's responsibilities to communities around the mines would be met (Mhlongo & Amponsah-Dacosta, 2016). It required mining companies to provide an environmental management programme and a rehabilitation plan (Swart, 2003). They had to provide details about post-mining provisions for environmental liabilities. But the requirements were difficult to enforce, and the act did not bring much change.

The Constitution (Republic of South Africa, 1996a) laid the foundation for a plethora of post-apartheid policies by using the concept of sustainability and for future mining legislation. The Mine Health and Safety Act (Republic of South Africa, 1996b) addressed the high fatality rate in the mining industry. Several environment-related statutes followed. The National Water Act (Republic of South Africa, 1998b) regulates water use and water quality management. Mine companies required a licence for water use and had to comply with water quality standards and submit a water management plan for their environmental impact assessment. The National Environmental Management Act (Republic of South Africa, 1998a) provided a framework for environmental management that included the environmental impacts of mining. This legislation also required mining companies to rehabilitate the mine's surroundings at closure. The Nuclear Energy Act (Republic of South Africa, 1999) introduced regulations for managing radioactive materials, including those in the mining industry. The Minerals, Petroleum and Resources Development Act (MPRDA) (Republic of South Africa, 2002) established a regulatory framework for managing mineral resources in the country throughout the mining lifecycle (Swart, 2003). This act is the primary legislation relating to the country's mineral resources, although it should be read in conjunction with other legislation (e.g. water and environmental laws). The MPRDA aimed to ensure that mining companies would take responsibility for local community growth and development (Mpanza et al., 2021). The government amended the MPRDA in 2002 to include measures for managing environmental impacts and financial provisions for rehabilitation.

The National Environmental Management: Air Quality Act (Republic of South Africa, 2004) and the National Environmental Management: Waste Act (Republic of South Africa, 2008c) regulate the management of these issues by all industries, including the mining sector. These regulations tackle the environmental consequences of mining activities and promote clean mining practices.

Despite all the legislation discussed earlier, South Africa faces a serious problem of abandoned mines. Fourie and Brent (2006) argue that South Africa has adequate policies on mine closure and that this extends to the social aspects of closure. However, Adler et al. (2007) contend that mine legislation has not been effectively implemented because of the fragmentation of responsibilities between different stakeholders. Mine closure legislation falls within the scope of environmental, water, waste and pollution laws. As a result, all levels of government are responsible for enforcing mining laws and regulations. Coordinating these responsibilities has been difficult. Consequently, mine tailings and dumps are mismanaged and abandoned (Mpanza et al., 2021). Swart (2003) says mine closure is unsuccessful because so many closure scenarios are possible, such as abandonment or dereliction, illegal temporary closure, temporary closure in line with the legislation or closure with a *closure* certificate. Not much has changed since this observation. Watson and Olalde (2019, p. 639) note the slow progress with improving the mine closure system, which they say 'as implemented in South Africa is largely ineffective'.

3.3 The Draft National Mine Closure Strategy (2021)

The Draft National Mine Closure Strategy was released in 2021, nearly 20 years after the government promulgated the MPRDA. As we noted earlier, the MPRDA established a regulatory framework for a mine's lifecycle, yet formal mine closure seldom occurs.

3.3.1 Content and objectives

The Draft Strategy sets the following five objectives (DMRE, 2021, p. 3):

- Manage the closure of mines in an integrated and sustainable manner, ensuring that the mines work together to obtain self-sustaining ecosystems after closure;
- Ensure that the mines do not impact negatively on the livelihood of adjacent mines;
- Promote a strategic approach to managing water at mining and mineral processing sites so that it is efficiently managed and valued, and develop a post-closure mine water strategy;
- Make provision for post-closure stewardship and socioeconomic sustainability, and monitor closure plans;
- Integrate environmental management and related closure activities with socio-economic interventions and align these with a post-closure economy by rationalising spending on environmental management programmes, social and labour plans, and corporate social responsibility (CSR) by reducing duplication of effort and spending.

These objectives are a step in the right direction, but they are very optimistic. However, integrating environmental and social aspects is an important contribution to mine closure thinking. This link is often assumed in mine closure strategies but seldom finds practical application. The emphasis on integration is crucial because environmental aspects can easily dominate.

In addition to the objectives, the Draft Strategy lists focus areas that include planning for closure beyond environmental rehabilitation, namely closure planning throughout a mine's lifecycle, a regional approach to closure, collaborative regional development and the identification of mine closure regions. The regional approach to closure and the re-emphasis on integrating socioeconomic and environmental aspects are important. Yet the regional closure approach might be difficult to achieve, and the role of regional economies might receive little attention. The Draft Strategy proposes an implementation plan consisting of the following:

- Economic succession planning and the development of a sustainable post-mining economy;
- Economic benchmarking;
- Planning a diverse post-mining economy;
- Integrating socioeconomic activities with mine environmental management and closure;
- Tailoring closure plans to various categories of mines;
- Peer review, monitoring and evaluation.

Overall, these ideas are sound, but the Draft Strategy is unclear on what they entail. For example, the term 'economic benchmarking' could be important, but no details are provided.

3.3.2 Critical assessment

Despite being a step in the right direction, we think the Draft Strategy has several shortcomings.

First, the Draft Strategy assumes that developing post-mining economies is easy and can be planned. Though such development is not impossible, the assumption is highly problematic for many reasons. Although lifecycle planning is common in the mining industry, it is hard for a mining company to think about the consequences of closure while trying to deal with the issues of opening a mine. Mine lifecycle planning is absent in communities and organisations, and the prospect of prosperity blinds them to the necessity of thinking about closure. Mining investments change mining areas rapidly, so managing the disruption associated with mine development is challenging.

Although long-term planning is possible, it requires information and local-level capacity, which is often lacking (Marais, 2023), and mines are often not transparent about their planning (Oranje & Nel, 2022). Long-term planning at the community level requires starting with the same baseline information as the mining companies. But the companies seldom share this information because it is business information and they do not want it made public. By making a single large financial investment to extract resources, a mining company changes the social environment. When mining communities plan, their planners tend to think that the same scale of investment will be available for economic diversification. This is not the case.

Economic diversification often depends on a new economic sector's relatedness to a previous sector, but there are not many economic activities related to mining. Very few global examples exist, and where they do, they are often the exception to the rule. Evolutionary economic geography points out that economic diversification processes are usually slow to evolve and depend on an existing space economy at a specific time. Planning economic diversification by thinking like a mining company, planners fail to notice these factors. Focusing on a grand alternative to mining, they assume that a diverse economy will automatically follow mine closure.

Second, the Draft Strategy fails to frame the consequences of mine closure in a systematic way. It makes little reference to how mining companies and local communities should respond when mines open or how mining affects local communities at closure. Many communities and mines make large-scale social, infrastructural and economic investments at the start of mining operations. To obtain a social licence to operate, mining companies have to show how communities will benefit from the mine. However, when a mine closes, the long-term dependencies that these investments create could become stumbling blocks. A good strategy should take this risk into account. Most of the Draft Strategy's references are to economic issues, with social aspects in the background. For example, in referring to water and land rehabilitation after closure, it makes little mention of the social aspects. The link between the two requires attention. It also fails to recognise that crime increases when a mine closes (Marais, Ndaguba et al., 2022). Theft of infrastructure by criminal syndicates is common. This inhibits further economic development. A systematic understanding of the nature and scale of social issues would go a long way towards tightening up the policy response to mine closure.

Third, the Draft Strategy assumes that plentiful local capability will be available to deal with the social aspects of mine closure. On the one hand, this expectation is legitimate, and localities should be the main respondents to mine closure. On the other hand, the scale of the social consequences might mean that dealing with them locally is impossible. A good strategy for mine closure should distinguish between local and other levels of responsibility. The national government is far too ready to cascade these responsibilities down to municipalities and communities. Most of South Africa's mining towns have local planning capacity constraints. Very few have the capacity to deal with the complexities of both mining growth and closure.

Fourth, the Draft Strategy fails to develop practical ways to help communities and their organisations plan for mine closure. Part of the problem is that the country's planning system is rooted in the new public management paradigm (Harrison, 2001). Consequently, plan-making and steering are prioritised over creating economic links. This often means an overly linear approach to economic planning, focusing on only one economic option. The result is simplistic assumptions about the future, ignoring the possibility of closure and banking on continued growth. Scenario planning, with different scenarios for growth and decline, could help solve this problem.

Sixth, like most strategy documents, the Draft Strategy takes no notice of what ordinary people do to mitigate the risk of closure, such as staying mobile in order to leave the area when the mine declines (Marais, 2023). Renting a house rather than owning it means that miners will not have assets in a town where their economic future is vulnerable. Failing to consider this, the Draft Strategy supports the current mine housing policy, which emphasises homeownership. It thus reinforces communities' risky long-term dependence on mining.

Finally, the Draft Strategy does not support the closure of defunct mining towns and the rehabilitation of that land, as is done with a defunct mine. Instead, it supports government policy that promotes open towns and democratic governance. Open towns were intended to reduce the role of mining companies in managing towns and cities. Although this approach by the government does have merit, it is assumed that the prosperity created by mining will continue. This seldom happens. We believe town closure or appropriate planning for the decline should be considered and could lay the foundation for future growth, but the Draft National Mine Closure Strategy needs to consider these options.

3.4 Conclusion

Legislation over the past 65 years has been inadequate to ensure appropriate mine closure in South Africa. It has not led to environmental rehabilitation being enforced, and it has largely ignored the social aspects of closure. The Minerals Act (50 of 1991) was the first solid attempt to legislate mine closure. The South African Constitution (Republic of South Africa, 1996a) emphasised sustainability and laid the foundation for numerous pieces of legislation. But proper mine closure remains elusive, despite a range of very good legislation since 1996. The lack of coordination and the turf wars between government departments compound the problem. The social aspects of mine closure remain poorly integrated into policy and practice. It is in this context that the government released the 2021 Draft National Mine Closure Strategy.

It is well known that mineral resources can be a curse, bringing social problems along with the wealth. Mine closure does not end the problems but increases them. Failing to imagine a time without mining, communities drift into dependence on the mine. Closure brings social disruption that hinders development. A mine closure strategy that does not take this into account will make ill-judged assumptions about the ease of creating a post-mining economy.

The expectation that mines and governments must create post-mining economies could easily develop into a fixation on economic diversification. The question needs to be asked whether such diversification is possible in South Africa's current space economy. Peripherally located mining areas will find economic diversification particularly difficult. It may be wiser to abandon the grand plan of post-mining prosperity and instead consider closing mining towns that are facing a bleak future. This would save money and make it possible to rehabilitate the land. The global experience has shown how difficult it is for a town to find a new path after mining because of the scarcity of related non-mining industries. We hope that future policy and legislation will consider the many urgent mine closure issues we have raised in this chapter.

References

Adler, R., Claassen, M., Godfrey, L. & Turton, R., 2007. Water, mining and waste: An historical and economic perspective on conflict management in South Africa. *Economics of Peace and Security Journal*, 2(2), pp. 367–372.

CSIR (Council for Scientific and Industrial Research), 2019. A CSIR perspective on South Africa's post-mining landscape. CSIR Report No. 16928. Pretoria: CSIR.

DMRE, 2021. Draft National Mine Closure Strategy, 2021. Government Gazette, Report no. 44607, 21 May.

Fourie, A. & Brent, A., 2006. A project-based Mine Closure Model (MCM) for sustainable asset life cycle management. *Journal of Cleaner Production*, 12–13, pp. 1085–1095.

Freudenburg, W.K.K., 1982. Increasing the impact of sociology on social impact assessment: Toward ending the inattention. *American Sociology*, 17, pp. 71–80.

Harrison, P., 2001. The genealogy of South Africa's integrated development plan. *Third World Planning Review*, 23(2), pp. 175–193.

Jones, M. & Mayzer, R., 2021. A fifth wave? The changing nature of boomtown research. *The Extractive Industries and Society*, 8(2), p. 100893.

Marais, L., 2023. *The Social Impacts of Mine Closure in South Africa: Housing Policy and Place Attachment*. London: Routledge.

Marais, L., Cloete, J. & Lenka, M., 2022. The plight of mining cities in South Africa: Planning for growth and closure. *Cities*, p. 103965.

Marais, L., Ndaguba, E., Mmbadi., E., Cloete, J. & Lenka, M., 2022. Mine closure, social disruption and crime in South Africa. *The Geographical Journal*, 188(3), pp. 383–400.

Mhlongo, S. & Amponsah-Dacosta, F., 2016. A review of problems and solutions of abandoned mines in South Africa. *International Journal of Mining, Reclamation and Environment*, 30(4), pp. 279–294.

Mpanza, M., Adam, E. & Moolla, R., 2021. A critical review of the impact of South Africa's mine closure policy and the winding-up process of mining companies. *The Journal for Transdisciplinary Research in Africa*, 17(1), a985.

Oranje, M. & Nel, V., 2022. Planning in the dark. In Marais, L., Burger, P., Campbell, M., Denoon-Stevens S. & van Rooyen, D. (eds.), *Coal and Energy in South Africa: Considering a Just Transition*. Edinburgh: Edinburgh University Press.

Swart, E., 2003. The South African legislative framework for mine closure. *The Journal of the Southern Africa Institute of Mining and Metallurgy*, October, pp. 489–492.

Watson, I & Olalde, M., 2019. The state of mine closure in South Africa – What the numbers say. *Journal of the Southern African Institute of Mining and Metallurgy*, 119(7), pp. 639–645.

Acts and policy documents referred to in the chapter

Department of Mineral Resources and Energy (DMRE). 2021. Draft National Mine Closure Strategy, 2021. Government Gazette, Report no. 44607, 21 May.

Republic of South Africa, 1965. Atmospheric Pollution Prevention Act (No. 45 of 1965), Government Gazette.

Republic of South Africa. 1967. Mining Rights Act (No. 20 of 1967), National Gazette, Vol. 27, Report no. 434, Cape ++Town.

Republic of South Africa, 1975. Fanie Botha Accord, 1975, Chamber of Mines.

Republic of South Africa. 1996a. The Constitution of the Republic of South Africa (Act No. 108 of 1996), Government Gazette.

Republic of South Africa. 1996b. Mine Health and Safety Act (No. 29 of 1996), Government Gazette, No. 17242, Notice No. 96, Cape Town.
Republic of South Africa. 1991. Minerals Act (No. 50 of 1991), Government Gazette, Vol. 311, Report no. 13253, Cape Town.
Republic of South Africa. 1998a. National Environmental Management Act (No. 107 of 1998). Government Gazette, Report no. 19519, 1540. Cape Town.
Republic of South Africa, 1998b. National Environmental Management: Waste Act (No. 59 of 2008), Government Gazette, Vol. 525, Report no. 32000, Cape Town.
Republic of South Africa. 1998c. National Water Act (No. 36 of 1998), Government Gazette, Vol. 398, Report no. 19182, Cape Town.
Republic of South Africa. 1999. Nuclear Energy Act (No. 46 of 1999), Government Gazette, Vol. 414, Report no. 20759, Cape Town.
Republic of South Africa. 2002. Mineral and Petroleum Resources Development Act (MPRDA) (No. 28 of 2002), Government Gazette, Vol. 448, Report no. 23922, Cape Town.
Republic of South Africa. 2004. National Environmental Management: Air Quality Act (No. 39 of 2004), Government Gazette, Report no. 27318, 476. Cape Town.
Union of South Africa. 1956. Mines and Works Act (No. 27 of 1956), Printed in the Union of South Africa by Cape Times Ltd, Parow.

4 A scoping review of the literature on mine closure

Anita Venter, Job Gbadegesin and David van Wyk

4.1 Introduction

Mining continues to generate substantial employment and serves as a source of livelihood for at least 10% of South Africa's population. Mine closure affects those dependent on mines directly or indirectly. Economic and environmental considerations dominate the literature, with little attention being paid to long-term social problems. Since the 1990s, the mine closure literature has had much to say about the processes of mine closure. Many closure frameworks come from mining companies, governments and international organisations like the World Bank. They tend to ignore social and community perspectives. They seldom account for the social cost, which can affect communities for many years after the closure. Even with closure plans, mining leaves a legacy of problems. Sometimes mining companies do not follow any plans but simply abandon the mine. Some partly close it, ceasing to mine without formally closing the mine and rehabilitating the area. This invites illegal activities. Some companies revive old mines, but these reopened mines rarely have the institutional memory of how to mitigate long-term social consequences.

Mines may close because of resource depletion or for other reasons, such as structural changes in the national economy (Strambo et al., 2019). Beckett et al. (2020) say mine closure is the longest and most complex phase of the mining cycle, with environmental, economic, social and regulatory consequences.

Determining the social impact depends on factors such as the geography and history of the area, the type of minerals mined, the type of mining, how long the mine has been in operation and the kind of waste generated. The method of mineral extraction (underground, surface, open pit and combinations of these) influences the resulting waste and the long-term impact on the environment and health (Haslam & Tanimoune, 2016), as does the type of mineral extracted (Mhlongo & Amponsah-Dacosta, 2016). Soil and water contamination can linger for decades or centuries after extraction, and waste from underground mines, although these mines look small on the surface, can affect communities far from the site (Phillips, 2016; Horowitz et al., 2018).

This chapter's review of the mine closure literature shows that the stress is often on how closure *should* be done (Faizuldayeva, 2016), but case studies of the

DOI: 10.324/9781003403326-4

social impacts after closure are scarce. Some plans and models incorporate socio-economic planning as a mere technical exercise of the mine life cycle planning while overlooking the perspectives and voices of local communities (Bainton & Holcombe, 2018).

We used the terms 'mine closure', 'social' and 'social impact' to search Google Scholar for relevant literature. This yielded 1,279 results, consisting of peer-reviewed articles, books, conference proceedings, postgraduate studies, reports and popular articles, about half of them technical. We narrowed this down to about 600 sources that included social aspects. Most of this literature is governance-related, providing and critically discussing models and frameworks for formal mine closure processes. The social themes are linked to mine life cycles, geopolitics, bio-regional dynamics and the type of mining. Some related themes are health implications, uniqueness of place, community dynamics, and the institutional and governance principles of the country investigated. Some sources consider the power dynamics that influence the social implications, particularly where there is an imbalance between developed and developing countries. Many sources consider the consequences of closure for the physical environment and how this affects communities. The reference lists of the articles in this database yielded some supplementary sources. We integrated our own knowledge of the topic and incorporated critical perspectives. We considered in particular the interdependence of ecological systems and human communities in resource extraction processes (Du Plessis & Brandon, 2015).

4.2 Mine closure legislation

Formal mine closure practices were rare until the 1990s. Since then, governments and mining companies have become more responsible about closure plans. Global mine closure regulation is a relatively new concept. Before the 2000s, only a handful of governments enacted mine closure laws, primarily concentrating on mitigating environmental damage and advancing economic agendas. However, government agencies, mine operators and experts from various disciplines seldom share a unified vision regarding mine closure (Owen & Kemp, 2017; Bainton & Holcombe, 2018). Social consequences appear to be a secondary concern in mine closure plans and seldom feature on balance sheets (Van Eeden et al., 2009; Bainton & Holcombe, 2018). The social costs remain poorly understood and are often underestimated (Watson & Olalde, 2019). International regulations require mine sites to be restored to a safe, non-polluting state that can support acceptable post-mining living conditions. While there are national and international recommendations for mine rehabilitation and closure, guidance on how to set quantifiable and attainable criteria for social rehabilitation is fragmented (Bainton & Holcombe, 2018). We found abundant models or frameworks for achieving successful closure, a few examples being Perman (2003), Du Plessis and Brent (2006), Sullivan and Amacher (2009), Mishra et al. (2012), Paricheh and Osanloo (2017) and Oliveira et al. (2020).

The International Council for Mining and Metals' good practice guide (ICMM, 2019) identifies distinct and context-specific criteria for measuring the social costs of mine closure. Consequently, global mining companies are now expected to respond to the numerous scientific publications critical of the social and ecological problems associated with mine closure. For example, the World Bank's mine closure 'toolbox' (2021) says that long-term closure and reclamation can only succeed if all physical structures are designed to eliminate potential hazards to humans, wildlife (including aquatic life) and environmental safety for 1,000 years. This seems like an impossibly tall order.

Despite many policies to regulate mine closures, it is not clear who should be held accountable. Failed attempts at closing mines with acceptable and sustainable social, economic and environmental results for communities can be attributed to institutional problems, financial difficulties, inadequate community participation and insufficiently detailed closure plans (Van Druten & Bekker, 2017).

4.3 Abandoned mines

Mines have been abandoned since the dawn of mining. They remain a big concern globally. An abandoned mine is one that does not have a clear owner responsible for rehabilitation after formal operations end. It is a public liability as measures are seldom put in place to contain or repair the damage (World Bank, 2021; Marais et al., 2021). The number of abandoned mines is staggering. Mackasey (2000) reported two decades ago that Canada had 10,000. China has at least 20,000 (Du et al., 2022). The United Kingdom today has more than 2,000. It is not always clear who should take responsibility for the thousands of abandoned mines that are still polluting the environment. There is little evidence that governments and mining companies are engaging with the problem (Owen & Kemp, 2017; Watson & Olalde, 2019; Marais et al., 2021). Abandoned mines can affect communities decades after operations have ceased (Gordon, 2015; World Bank, 2021). They pose an immediate danger if there are underground fires, open pits or shafts that are not properly sealed (Limpitlaw et al., 2005; Sango et al., 2006).

4.4 Social costs

The social costs of mine closure include socio-economic, health and cultural costs, as well as institutional costs (such as local governments lacking capacity to take over from the mining companies). The costs are often the result of dependencies created by the mines and the difficulty of making the transition to a different local economy (Sango et al., 2006; McDonald et al., 2012; Bainton & Holcombe; 2018).

Most regulatory documents are technical and provide remedial approaches rather than comprehensive measurement frameworks. The current short-term focus is on technical exercises that are unlikely to satisfy communities' long-term needs. Issues covered by social cost mine closure frameworks are stakeholders' engagement prior to closure; ecological, land-use and tenure concerns; risk assessments; quality and accountability monitoring; and technological, special purpose and integrated approaches. Mining companies are inclined to focus on costs directly related to operations and managerial aspects of social relations that they can control, whereas governments prefer to calculate costs and benefits to the country. The frameworks tend to exclude community voices and NGOs in assessing social costs and their impacts on society (Bainton & Holcombe, 2018).

4.4.1 Health

There is substantial literature on the health impacts of closed and abandoned mines worldwide. We review it here because poor health has social effects. Health concerns related to mining operations are often overlooked in conventional social cost calculations. Pollution of water and agricultural soil gets into the food chain (Finucane, 2008; Mhlongo & Amponsah-Dacosta, 2016; Ackermann et al., 2018). Acid mine drainage can spread further during heavy rains, polluting water resources. Illnesses from exposure to radioactive materials and noxious minerals can be severe, either through working on a mine or living close to one. Dust and airborne particles from mine tailings may contain silicates and toxic substances that cause respiratory conditions (Utembe et al., 2015). Deformities due to toxic exposure during pregnancy have been reported (Horowitz et al., 2018).

Physical dangers, such as tailings dam collapses, may cause loss of property and serious injuries or death (Edraki et al., 2014). Abandoned mines may leave unsafe physical infrastructure that can lead to accidents and injuries (Sango et al., 2006; Ackermann et al., 2018). Mining company documents may list injuries and deaths that occur on the mines' premises when a mine is in 'care and maintenance' but fail to fully account for the broader chronic diseases that affect communities long after the mine is closed or abandoned (Winde & Sandham, 2004; Nkosi et al., 2015; Parshley & MacCallum, 2016; Kengni & Nkosi, 2023).

Psychological problems such as stress, hopelessness, despair, emotional trauma and other mental health disorders are mentioned by several authors (Sango et al., 2006; D'Silva & Norman, 2015; Mactaggart et al., 2016; Ackermann et al., 2018; Siyongwana & Shabalala, 2019). The disruption of social networks by the outmigration of people seeking work elsewhere and the resulting isolation, especially in the remote areas where many mines are situated, can damage mental health.

Most mines provide medical benefits to employees and their families, but these generally cease with closure (Ackermann et al., 2018). The mines may also invest in health infrastructure (clinics or hospitals) for workers and the community, but after closure, there is rarely funding to keep them operational. For rural communities, this can mean curtailed services or long distances to travel for healthcare (Sango et al., 2006; Mactaggart et al., 2016; Marais, 2023).

4.4.2 Poverty

Although mines usually pay retrenchment benefits, this is very small compensation for loss of income. Mine closure gradually depletes mining communities' livelihood assets, reducing their living standards (Ackermann et al., 2018). Many former miners lack the skills to find work elsewhere or cannot afford the travel expenses to seek jobs elsewhere. Mining economies generally cannot diversify after closure

(Andrews-Speed et al., 2005), which means there will not be many new jobs. Dependence on mining can cause innovative plans for alternative economies and livelihoods to be rejected (Wirth & Harfst, 2012). Older miners may refuse to learn new skills (Strambo et al., 2019). Contaminated land and water may not be usable for crops or livestock, which means farming is not an option (Sango et al., 2006; George-Laurentiu et al., 2016). Lack of purchasing power in the community may force businesses to close, leaving the owners unemployed. Lack of income means lack of food, leading to undernutrition and malnutrition. Children drop out of school because parents cannot afford the fees (Sango et al., 2006; Siyongwana & Shabalala, 2019; Ntema et al., 2023). Loss of mine income reduces municipal revenue, and the municipality struggles to provide services (Ntema et al., 2023; Siyongwana & Shabalala, 2019). House maintenance gets neglected, and increased poverty contributes to lower self-esteem and perceptions of powerlessness (Gamu et al., 2015).

4.4.3 Crime

Marais et al. (2022) have documented an increase in crime where mining is in decline. Sango et al. (2006) and Siyongwana and Shabalala (2019) reported similar findings. Axbard et al. (2016) found a decrease in crime in the initial phases of mining but a sharp increase at closure, in the form of theft, carjacking, trafficking narcotics and violent crime such as rape and murder.

4.5 Cultural identity

A community's cultural identity is made up of its norms and values. In small mining towns, the mine is the local community's hub and its 'most important economic, social and cultural matrix' (Wirth & Harfst, 2012, p. 10). Mine closure disrupts social networks and weakens them through the outmigration of skilled workers (Strambo et al., 2019).

Europe, the United States, Canada and Australia have many post-mining towns (Oakley, 2018). Over the years, some mining communities have developed their own deeply rooted cultural identity. Their accrued social and cultural capital can facilitate the town's regeneration. Towns with a strong culture shaped by a mining identity can be found in many countries, such as Ireland, Scandinavia, the United States, Canada, Australia and Chile (Devenin & Bianchi, 2019; Rossi et al., 2021).

Mining communities that have developed their own culture and sense of identity over time may strongly oppose mine closure. For them, mining continues to drive their identity, directly or indirectly. Living near the mine, they come to rely not only on the mine itself but also on the supporting infrastructure (Horowitz et al., 2018). Former mining towns that no longer have active mines often lose their mining identity and may turn to heritage preservation in the hope of attracting tourism (Rossi et al., 2021).

Abruptly closing relatively young mines can have social costs, including losing cultural identity even before the mining town fully shapes it. Often, closure is already a reality even in the boom phase of the mine, and miners who migrate lose their connection to their place of origin (Rossi et al., 2021). Mining communities where a mine is in the early stages of establishment and experiencing rapid cycles of booms and busts may have yet to achieve social stability and cultural identity and be more susceptible to socio-environmental conflicts before and after closure (Haslam & Tanimoune, 2016; Walter & Urkidi, 2017). Relatively young mines constantly transform an area's identity. In-migration leads to a fusion of cultures, cultures transforming and communities finding themselves between a past identity and a new identity. When the mine closes, the culture can be diminished or lost, particularly for the younger generation, who do not have a connection to their ancestral place of origin or to the mining industry as their parents and grandparents did (Rossi et al., 2021).

4.5.1 Cultural and community resistance to mining

Yet despite mines being abandoned worldwide, global mining companies continue exploitative extractions in developing countries where laws are less stringent and labour is cheaper. Latin America experienced a remarkable surge of 400% in mining investment, with an astounding escalation of 2000% in Peru (McDonell, 2015). Urkidi (2010) says Canada and the United States venture into developing countries largely to avoid mining restrictions in their countries.

A spin-off effect of community distress caused by mine closure is community resistance to opening mines in the first place. Worldwide environmental movements are expressing anti-mining concern and putting pressure on existing mines to close while opposing the opening of new mines (Urkidi, 2010; Walter & Urkidi, 2017). The notable surge in mining activities in South America has generated a distinct body of literature exploring social resistance, pressure to close mines and resistance to cultural alienation. Conflicts arise over allegations of resource exploitation by multinational companies and environmental degradation in regions where culture and nature are closely connected (Urkidi, 2010; Haslam & Tanimoune, 2016; Rossi et al., 2021). Urkidi (2010) and McDonell (2015) discuss the problems of mining expansion in South America and the pressure being brought on mines to close in favour of upholding indigenous knowledge and cultural values. Opposition to mining in Latin America is driven by concerns over environmental degradation, human rights, livelihoods and extreme weather events (Urkidi, 2010; Vargas, 2019).

4.5.2 Ecological resistance to uphold indigenous practices

Some studies point out the particular social costs of mining for Indigenous communities with a strongly nature-based culture. Large-scale mining profoundly transforms cultural landscapes (Haslam & Tanimoune, 2016). The consequences of mine closure for Indigenous communities are a topic of ongoing research (O'Faircheallaigh & Lawrence, 2019). Anti-mining groups complain about historical disregard for indigenous knowledge, disrespect for ancestral land, the creation of excessive reliance on mining activities and social issues arising from the loss of traditional livelihood practices in support of capitalist ideals (Horowitz et al., 2018). The same countries that produce good practice guides for mine closure are exploiting new territories in Latin America and Africa. These good practices need to be implemented to avoid the damage that can be done to communities by mine closure.

4.6 Proposals for dealing with the social impacts of mine closure

The proposals in the literature for countering detrimental impacts can be grouped into two kinds: comprehensive planning for mine closure and community engagement during the planning and closure process. Many authors acknowledge the shortcomings of the social licence to operate, usually as part of an environmental impact assessment, when it comes to closure planning (Finucane, 2008; Lockie et al., 2009; Lamb & Coakes, 2012; Strambo et al., 2019). The more detailed plans that must be implemented towards closure must address social, economic and environmental aspects. Building new economies is vital to the process (Kemp, 2010). Community involvement in the planning of the closure is critical. It should entail collaborative planning and not merely top-down tokenism (Arnstein, 1969; Healey, 2003). Among the authors stressing community involvement are Lamb and Coakes (2012), Costa (2015), Van Druten and Bekker (2017) and Monosky and Keeling (2021). Good governance by local, provincial and national governments to support communities in distress has often been lacking (Wirth & Harfst, 2012; Siyongwana & Shabalala, 2019).

4.7 Conclusion

The social costs mentioned in the mine closure literature, whether academic or from the mining industry, are multi-dimensional and not universally quantifiable. They must be seen in the context of the structural legacy of mining. Most social costs are related to health, livelihoods and cultural identity. Mine-related pollution affects the health of the environment and, hence, crops, livestock and humans. Mine closure often devastates the economy, leading to psychological problems and crime. Mining and cultural identities are inherently linked, with different identities evolving over time and some communities developing a strong attachment to the heritage of mining. Indigenous communities strive to preserve cultural practices closely tied to the land through ecological resistance.

Although there is growing awareness of the social costs of mining, the proposals to deal with these costs are mostly limited to curtailing the environmental impacts, making more detailed plans for closure and encouraging community participation in the process. Monitoring compliance with the legislation is essential. Local and other levels of government should help communities develop new skills, engage in new livelihoods and build new economies. A major stumbling block, as discussed throughout this book, is the dependencies developed over the life of a mine that inhibit communities' social and economic survival and recovery.

References

Ackermann, M., Botha, D. & Van der Waldt, G., 2018. Potential socio-economic consequences of mine closure. *TD: The Journal for Transdisciplinary Research in Southern Africa*, 14(1), pp. 1–11.

Andrews-Speed, P., Ma, G., Shao, B. & Liao, C., 2005. Economic responses to the closure of small-scale coal mines in Chongqing, China. *Resources Policy*, 30(1), pp. 39–54.

Arnstein, S.R., 1969. A ladder of citizen participation. *Journal of the American Institute of Planners*, 35(4), pp. 216–224.

Axbard, S., Poulsen, J. & Tolonen, A., 2016. Extractive industries, production shocks and criminality: Evidence from a middle-income country. CDEP-CGEG Working paper 30, Center on Global Energy Policy. Columbia University, New York.

Bainton, N. & Holcombe, S., 2018. A critical review of the social aspects of mine closure. *Resources Policy*, 59, pp. 468–478.

Beckett, C., Dowdell, E., Monosky, M. & Keeling, A., 2020. Integrating socio-economic objectives for mine closure and remediation into impact assessment in Canada. *Draft Report for the SSHRC Knowledge Synthesis Grant: Informing Best Practices in Environmental and Impact Assessment.* https://research.library.mun.ca/14487/1/KSG%20Grant_Full%20Report_FinalDraft_Mar31_2020.pdf

Costa, S. 2015. Social impacts of mine closure: Engaging employees and host communities in planning for closure. *British Columbia Mine Reclamation Symposium, Norman B. Keevil Institute of Mining Engineering, University of British Columbia, Vancouver.* https://open.library.ubc.ca/soa/cIRcle/collections/59367/items/1.0305870

Devenin, V. & Bianchi, C., 2019. Characterising a mining space: Analysis from case studies in Chile and Australia. *Resources Policy*, 63, 101402.

D'Silva, S., & Norman, P., 2015. Impacts of mine closure in Doncaster: An index of social stress. *Radical Statistics*, 112, pp. 23–33.

Du, K., Xie, J., Khandelwal, M. & Zhou, J., 2022. Utilisation methods and practice of abandoned mines and related rock mechanics under the ecological and double carbon strategy in China: A comprehensive review. *Minerals*, 12(9), 1065.

Du Plessis, A. & Brent, A.C., 2006. Development of a risk-based calculation model of mine closure cost. *Journal of the South African Industry of Mining and Metallurgy*, 106, pp. 443–450.

Du Plessis, C., & Brandon, P., 2015. An ecological worldview as basis for a regenerative sustainability paradigm for the built environment. *Journal of Cleaner Production*, 109, pp. 53–61.

Edraki, M., Baumgartl, T., Manlapig, E., Bradshaw, D., Franks, D.M. & Moran, C.J., 2014. Designing mine tailings for better environmental, social and economic outcomes: A review of alternative approaches. *Journal of Cleaner Production*, 84, pp. 411–420. https://doi.org/10.1016/j.jclepro.2014.04.079

Faizuldayeva, Z., 2016. A comparative study of regulatory approaches to mine closure with a special emphasis on the current situation in the former Soviet Union. Mine Closure 2016, Proceedings of the 11th International Conference on Mine Closure. Australian Centre for Geomechanics, University of Western Australia, Crawley. https://papers.acg.uwa.edu.au/p/1608_25_Faizuldayeva/

Finucane, S., 2008. Thinking about the end before you start: Integrating mine closure planning into feasibility studies and environmental and social impact assessment: Mine Closure 2008. Proceedings of the 3rd International Seminar on Mine Closure, Australian Centre for Geomechanics University of Western Australia, Crawley. https://papers.acg.uwa.edu.au/p/852_16_Finucane/

Gamu, J., Le Billon, P. & Spiegel, S., 2015. Extractive industries and poverty: A review of recent findings and linkage mechanisms. *The Extractive Industries and Society*, 2(1), pp. 162–176. https://doi.org/10.1016/j.exis.2014.11.001

George-Laurentiu, M., Florentina-Cristina, M. & Andreea-Loreta, C., 2016. The assessment of social and economic impacts associated to an abandoned mining site case study: Ciudanovita (Romania). *Procedia Environmental Sciences*, 32, pp. 420–430.

Gordon, S.M., 2015. Narratives unearthed, or, how an abandoned mine doesn't really abandon you. In Keeling, A. & Sandlos, J. (eds.), *Mining and Communities in Northern Canada: History, Politics, and Memory*. Calgary: University of Calgary Press, pp. 59–86.

Haslam, P.A. & Tanimoune, N.A., 2016. The determinants of social conflict in the Latin American mining sector: New evidence with quantitative data. *World Development*, 78, pp. 401–419.

Healey, P., 2003. Collaborative planning in perspective. *Planning Theory*, 2(2), pp. 101–123.

Horowitz, L.S., Keeling, A., Lévesque, F., Rodon, T., Schott, S. & Thériault, S., 2018. Indigenous peoples' relationships to large-scale mining in post/colonial contexts: Toward multidisciplinary comparative perspectives. *The Extractive Industries and Society*, 5(3), pp. 404–414.

ICMM (International Council for Mining and Metals), 2019. *Integrated Mine Closure: Good Practice Guide*. London: ICMM.

Kemp, D., 2010. Mining and community development: Problems and possibilities of local-level practice. *Community Development Journal*, 45(2), pp. 198–218.

Kengni, B. & Nkosi, V., 2023. Analysis of the current legal framework protecting the health of communities near gold mine tailings in South Africa. *Southern African Public Law*, 37(2). https://doi.org/10.25159/2522-6800/12919

Lamb, K. & Coakes, S., 2012. Effective social planning for mine closure. Mine Closure 2012: Proceedings of the 7th International Conference on Mine Closure. Australian Centre for Geomechanics, University of Western Australia, Crawley. https://papers.acg.uwa.edu.au/p/1208_53_Lamb/

Limpitlaw, D., Aken, M., Lodewijks, H. & Viljoen, J., 2005. Post-mining rehabilitation, land use and pollution at collieries in South Africa. Paper presented at Colloquium: Sustainable Development in the Life of Coal Mining, South African Institute of Mining and Metallurgy, Boksburg, 13 July.

Lockie, S., Franettovich, M., Petkova-Timmer, V., Rolfe, J. & Ivanova, G., 2009. Coal mining and the resource community cycle: A longitudinal assessment of the social impacts of the Coppabella coal mine. *Environmental Impact Assessment Review*, 29(5), pp. 330–339. https://doi.org/10.1016/j.eiar.2009.01.008

Mackasey, W.O., 2000. *Abandoned Mines in Canada*. WOM Geological Associates, Ontario. https://miningwatch.ca/sites/default/files/mackasey_abandoned_mines.pdf

Mactaggart, F., McDermott, L., Tynan, A. & Gericke, C., 2016. Examining health and well-being outcomes associated with mining activity in rural communities of high-income countries: A systematic review. *Australian Journal of Rural Health*, 24(4), pp. 230–237. https://doi.org/10.1111/ajr.12285

Marais, L., 2023. *The Social Impacts of Mine Closure in South Africa: Housing Policy and Place Attachment*. London: Routledge.

Marais, L., Burger, P., Campbell, M., Denoon-Stevens, S.P. & Van Rooyen, D., 2021. *Coal and Energy in South Africa: Considering a Just Transition*. Edinburgh: Edinburgh University Press.

Marais, L., Ndaguba, E., Mmbadi, E., Cloete, J. & Lenka, M., 2022. Mine closure, social disruption, and crime in South Africa. *The Geographical Journal*, 188(3), pp. 383–400. https://doi.org/10.1111/geoj.12430

McDonald, P., Mayes, R. & Pini, B., 2012. Mining work, family and community: A spatially-oriented approach to the impact of the Ravensthorpe Nickel Mine closure in remote Australia. *Journal of Industrial Relations*, 54(1), pp. 22–40. https://doi.org/10.1177/0022185611432382

McDonell, E., 2015. The co-constitution of neoliberalism, extractive industries, and indigeneity: Anti-mining protests in Puno, Peru. *The Extractive Industries and Society*, 2(1), pp. 11–123.

Mhlongo, S.E. & Amponsah-Dacosta, F., 2016. A review of problems and solutions of abandoned mines in South Africa. *International Journal of Mining, Reclamation and Environment*, 30(4), pp. 279–294. https://doi.org/10.1080/17480930.2015.1044046

Mishra, S.K., Hitzhusen, F.J., Sohngen, B.L. & Guldmann, J.M., 2012. Costs of abandoned coal mine reclamation and associated recreation benefits in Ohio. *Journal of Environmental Management*, 100, pp. 52–58.

Monosky, M. & Keeling, A. 2021. Social considerations in mine closure: Exploring policy and practice in Nunavik, Quebec. *Northern Review*, 52, pp. 29–60.

Nkosi, V., Wichmann, J. & Voyi, K., 2015. Chronic respiratory disease among the elderly in South Africa: Any association with proximity to mine dumps? *Environmental Health*, 14, pp. 1–8.

Ntema, J., Marais, L., Cloete, J. & Lenka, M., 2023. Mine closure in Matjhabeng in the Free State Goldfields, South Africa: Implications for households. *The Extractive Industries and Society*, 14, p. 101257.

Oakley, P., 2018. After mining: Contrived dereliction, dualistic time and the moment of rupture in the presentation of mining heritage. *The Extractive Industries and Society*, 5(2), pp. 274–280.

O'Faircheallaigh, C. & Lawrence, R., 2019. Mine closure and the aboriginal estate. *Australian Aboriginal Studies*, 1, pp. 65–82.

Oliveira, R., Zanella, A. & Camanho, A.S., 2020. A temporal progressive analysis of the social performance of mining firms based on a Malmquist index estimated with a benefit-of-the-doubt directional model. *Journal of Cleaner Production*, 267, p. 121807.

Owen, J.R. & Kemp, D., 2017. *Extractive Relations: Countervailing Power and the Global Mining Industry*. London: Routledge.

Paricheh, M. & Osanloo, M., 2017. A simulation-based framework for estimating probable open-pit mine closure time and cost. *Journal of Cleaner Production*, 167, pp. 337–345.

Parshley, J.V. & MacCallum, C.S., 2016. Attractive nuisances and wicked solutions. Mine Closure 2016: Proceedings of the 11th International Conference on Mine Closure, Australian Centre for Geomechanics, University of Western Australia, Crawley. https://papers.acg.uwa.edu.au/p/1608_15_Parshley/

Perman, R., 2003. *Natural Resource and Environmental Economics*. London: Pearson Education.

Phillips, J., 2016. Climate change and surface mining: A review of environment-human interactions & their spatial dynamics. *Applied Geography*, 74, pp. 95–108.

Rossi, M., Forget, M., Gunzburger, Y., Bergeron, K.M., Samper, A. & Camizuli, E., 2021. Trajectories of mining territories: An integrated and interdisciplinary concept to achieve sustainability. *The Extractive Industries and Society*, 8(1), pp. 1–7.

Sango, I., Taru, P., Mudzingwa, M., & Kuvarega, A., 2006. Social and biophysical impacts of Mhangura copper mine closure. *Journal of Sustainable Development*, 8(3), pp. 186–204.

Siyongwana, P.Q. & Shabalala, A., 2019. The socio-economic impacts of mine closure on local communities: Evidence from Mpumalanga province in South Africa. *GeoJournal*, 84(2), pp. 367–380. https://doi.org/10.1007/s10708-018-9864-5

Strambo, C., Aung, M. T. & Atteridge, A., 2019. Navigating coal mining closure and societal change: Learning from past cases of mining decline. Working paper, Stockholm Environment Institute, Stockholm.

Sullivan, J. & Amacher, G.S., 2009. The social costs of mineland restoration. *Land Economics*, 85(4), pp. 712–726.

Urkidi, L., 2010. A glocal environmental movement against gold mining: Pascua-Lama in Chile. *Ecological Economics*, 70(2), pp. 219–227.

Utembe, W., Faustman, E., Matatiele, P. & Gulumian, M., 2015. Hazards identified and the need for health risk assessment in the South African mining industry. *Human & Experimental Toxicology*, 34(12), pp. 1212–1221. https://doi.org/10.1177/0960327115600370

Van Druten, E. & Bekker, M.C., 2017. Towards an inclusive model to address unsuccessful mine closures in South Africa. *Journal of the Southern African Institute of Mining and Metallurgy*, 117(5), pp. 485–490.

Van Eeden, E.S., Liefferink, M. & Durand, J.F., 2009. Legal issues concerning mine closure and social responsibility on the West Rand. *TD: The Journal for Transdisciplinary Research in Southern Africa*, 5(1), pp. 51–71.

Vargas, G.A., 2019. Social mobilisation in Colombia's extractive industries, 2000–2015. *The Extractive Industries and Society*, 6(3), pp. 873–880.

Walter, M. & Urkidi, L., 2017. Community mining consultations in Latin America (2002–2012): The contested emergence of a hybrid institution for participation. *Geoforum*, 84, pp. 265–279.

Watson, I. & Olalde, M., 2019. The state of mine closure in South Africa – What the numbers say. *Journal of the Southern African Institute of Mining and Metallurgy*, 119(7), pp. 639–645.

Winde, F. & Sandham, L.A., 2004. Uranium pollution of South African streams: An overview of the situation in gold mining areas of the Witwatersrand, 61, pp. 131–149.

Wirth, P. & Harfst, J., 2012. *Challenges of Post-Mining Regions in Central Europe*. München: IÖR. www2.ioer.de/recherche/pdf/2012_wirth_challenges_of_post-mining_regions.pdf

World Bank, 2021. *Mine Closure: A Toolbox for Governments*. Washington, DC: World Bank.

5 Miners' lives after retrenchment

*Martina Kotzé, Boitumlo Joy Alec,
Mmboneni Magadzu and Jan Cloete*

5.1 Introduction

Mine downscaling and closure usually result in large-scale retrenchments and hardship for many stakeholders. The mines are often the only or most important economic activity keeping local mining communities afloat. In many places, especially in remote areas that lack effective local government structures, mining companies act as the local governing body. Dependence on the mining company means that mine closure and the subsequent job losses frequently lead to social disruption and leave the community traumatised (CDS, 2006; Coffie et al., 2021; Marais, 2013, 2023; Marais et al., 2005, 2022; Rixen & Blangy, 2016).

Mine closures have had negative implications worldwide, as has been found, for example in the United States (Nichols, 2013), India (Chandra, 2014), Romania, Russia and Ukraine (Haney & Shkaratan, 2003). But the repercussions in developing countries, where jobs are generally less secure and alternative socio-economic options are limited, may be more severe (Marais et al., 2005; Posel et al., 2021). In developing countries, in contrast to developed countries, mining revenue rose steeply in the decades between the 1960s and the 1980s, and in most southern African countries, the mining sector remains a major contributor to economic growth (IPPR, 2017). However, as most of these mines' lifespans are now ending, downscaling and eventual closure are unavoidable (CDS, 2006). In South Africa, the effects of mine closure are worsened by the abrupt desertion of some mines and the continual reselling of mines to companies that lack adequate resources to comply with closure regulations (Cornelissen et al., 2019). In many countries, retrenchment packages (Coffie et al., 2021), also referred to as 'redundancy allowance, separation or severance allowance' (Chandra, 2014, p. 238), are paid to help the victims of retrenchment bridge a period of unemployment, but this does not always happen in South Africa. And such packages assist retrenched workers only in the short term, leaving them, their families and their communities in crisis.

In this chapter, we investigate the effects of retrenchment on miners and their families in two case study areas affected by mine downscaling and closure. We look at how retrenchment took place, how it affected the miners, how they dealt with it and what alternative livelihoods were available to them. We found evidence

DOI: 10.324/9781003403326-5

of social disruption at both community and household levels and saw how people remained dependent on mine employment even after closure.

5.2 Retrenched: now what?

The effects of unemployment have been well researched in advanced, high-income countries, such as Finland (Martikainen et al., 2007), Germany (Pohlan, 2019) and Sweden (Mörk et al., 2020), but less in developing countries. We thus considered it important to investigate the effects of unemployment caused by retrenchment from the mines in South Africa, an upper-middle-income country.

Our first case study was conducted in the Free State Goldfields area in 2019 (see Chapter 6) and the second at Blyvooruitzicht gold mine on the West Rand (see Chapter 7) in early 2020.

Mining has declined rapidly in the Free State Goldfields over the past three decades (Marais, 2013; see Chapter 7), with many mines retrenching workers. Blyvooruitzicht, 80 km from Johannesburg, is a special case of a mine changing hands and workers being left in the lurch during the change. Most of its miners lived in mine hostels and mine housing in Blyvooruitzicht New Village, with the mining company providing essential infrastructure and services. In 1997, DRD Gold took over the mine but ran into financial difficulties and passed it to Village Main Reef, which began operational management of the mine despite pre-set conditions not having been fulfilled. Faced with the falling gold price and the cost of sustaining the mine or closing it properly, both companies disclaimed ownership of the mine and responsibility for it. The mine was placed under provisional liquidation in 2013, and mining ceased overnight. A property developer's offer to buy the houses in the village was rejected by the municipality, which refused to declare the village part of the Merafong City Local Municipality, as required by the developer. The mine was then occupied by illegal miners (known as *zama zamas*) to the detriment of the local community. The retrenched miners were given no notice of the closure, nor were they consulted. The Blyvooruitzicht community has been living in the vicinity of the deserted mine for years in conditions of extreme poverty, deprivation and fear of social decay. In 2015, the mine was purchased by Blyvoor Gold and was supposed to open again in April 2020, but at the time of the interviews, it was still not in operation (LHR & FIDH, 2016; Maeko, 2020a, 2020b).

The case studies investigated the effects of retrenchments on miners and their families. The participants were selected by non-probability sampling. There were 10 men and 5 women in the Goldfields case study and 13 men and 3 women in the Blyvooruitzicht case study. They had an average age of around 50 years. The interviewers asked questions like, 'What were the consequences for you and your family after losing this mining job?', 'What helped you make a living after mining?' or 'Tell us your story after losing your job at the mine'. The interview data were thematically analysed, and we use these themes to structure the findings reported here on their experiences, the effect on their families, and how they coped.

5.3 The plight of retrenched miners

5.3.1 Unemployed and penniless

Income loss is the most obvious and direct consequence of retrenchment and mine closures. Many participants in the Goldfields study considered that they had been 'dumped into poverty' despite receiving retrenchment packages. Victims of mine closure in Blyvooruitzicht were not fortunate. They all talked about how they had struggled due to a lack of income since the mine's provisional liquidation in 2013. They had received no retrenchment package or even outstanding wages for work already done when the mine suddenly closed. They were shocked to discover that the mine owner had disappeared while they were working underground. Those affected by the liquidation received notice that they would not receive any remuneration due until the new owner took charge of operations. One participant complained that he had not received money for his service but had been told by 'one of the union leaders' that 'a car is bought with its wheels', implying that the retrenched workers would be re-employed and paid by the new mine owner.

The prolonged period without income suffered by participants from both study areas meant that many had to depend on SASSA (South African Social Security Agency) grants for survival. In the Blyvooruitzicht case, the liquidators tried to sell some of the mine's assets while the retrenched miners waited for months and even years without being paid their wages or retrenchment packages (LHR & FIDH, 2016). Several people from the Goldfields area also depended on the Child Support Grant. One of them said, 'truly we are so struggling'. She said she depended on social grants for just two children because the eldest one had completed Grade 12 and did not qualify for the grant anymore.

How to pay off debt was a major problem for participants in both studies. One of the Blyvooruitzicht participants mentioned difficulty paying off a bank loan, and another was unable to keep up with mortgage repayments. Many Goldfields participants had to use their retrenchment packages to pay off debt and secure assets by paying off mortgages and other loans. This left them with little disposable income. One of them said he had used the retrenchment money to pay 'the debts that one accumulated during the working time'. He said, 'I had to pay all my accounts as there was no income anymore, and I was the only one supporting my family as the head of the house'.

It was clear that some participants were unable to manage their retrenchment benefits effectively. Some Goldfields participants said the mining companies had provided retrenchment letters for submission to creditors. These letters had not been helpful but only served to alert creditors to the miners' financial situation. Many participants hoped that the retrenchment letters would write off their debt and were disappointed to find they still had to pay their creditors. One said, 'the retrenchment letter given to us is not working when you show it to these stores'. Another understood, however, that the letter served 'to advise your creditors that you are not working, so you cannot pay your debts' and that it was 'a temporary arrangement when one is not working, but you need to pay your debts when you start working'.

5.3.2 Searching in vain for employment

Many participants were struggling to find a job. One Goldfields study participant said, 'I tried to find some jobs, but unfortunately I was always unsuccessful. I feel I was not lucky'.

Most said they were looking for jobs in mining, despite the scarcity of opportunities, as they had not worked outside the industry and planned to be miners for the rest of their working lives. The reasons they gave were their age, their skills being specific to the mining industry and their networks in the local mining labour market.

Participants in the Blyvooruitzicht study said they were waiting for the mine to reopen. They said the new mine owners had assured them that, despite their age, those who had lost their jobs would be the first to be employed when the mine reopened. They were nevertheless looking for other jobs since they knew the mines preferred to hire younger people.

At least two Blyvooruitzicht study participants alleged that there was corruption in job allocation. They felt the mines provided jobs to family members, friends or those paying bribes. They thought the job advertisements were a mere formality, as the employer already knew who would be appointed. One of them mentioned 'the corruption we are having here in South Africa'. She said, 'you must know somebody inside – you will hand in your CV, but if that CV does not go straight to the hands of the person who is hiring, forget about the job'.

Another Blyvooruitzicht participant thought that corrupt mine managers appointed non-South Africans because they could get away with paying them lower wages. This could cause conflict in the community as it leads to rivalry between foreign miners and unemployed South African miners. Xenophobia is not new in post-apartheid South Africa. It manifests in various forms, from simple intolerance and anti-immigrant attitudes to the xenophobic attacks that periodically flare up, as was seen in 1998, 2008, 2015 and 2019 in various parts of the country (Tella, 2016).

Six Blyvooruitzicht participants said they were affected by corrupt activities, and one said he had been invited to take part in criminal activities for an income:

> Even if I decide to take this money, it is going to kill me. It will kill me because I will be living on the tears of the people I worked with for more than 30 years. If you thought that I would take money for bribery and sell the people [probably meaning betray my principles and cheat people], I am not like that, sorry.

If the mine closure process is not handled effectively, many opportunities for corrupt activities present themselves, with negative implications for those who were retrenched as well as for the relevant communities. Comments like these suggest that this may have been the case at Blyvooruitzicht.

Self-employment is an option for retrenched people, but only four participants in the Goldfields study said they had tried to start up small businesses: buying and selling women's clothing, selling fruit and vegetables, and brickmaking. However, small businesses like this often do not generate enough income to survive, and many fail due to financial mismanagement, inadequate marketing strategies and an inability to collect payment. Our participants had not received training and lacked the experience to manage a small business. The mining industry does not foster entrepreneurial acumen. Some stories we heard were:

> I tried to do business selling clothes. The problem is that people will take clothes with credit, but you struggle when they have to pay. So it was not good business.
>
> Our project could not go further as hoped because the marketing strategy was inferior. We supplied the bricks for the construction company. Our company consisted of other ex-miners. We were in fact not making enough money from this company.
>
> There was a fight among us concerning the management of funds.

The participants' difficulty in finding jobs outside the mining sector and their limited success in entrepreneurial activities may be blamed on the government's inability to see the path dependence caused by mining employment. Building on a mining job to diversify the economy is difficult because the available skills are often only related to mining. The emphasis on mining as a tool for development and a misplaced belief that the country can continue to rely on mining as in the past have blinded the government to any solutions other than diversifying mining towns' economies. This leads to a crisis when mines decline because the much-needed economic diversification is difficult to achieve. Mining's lack of 'relatedness' means that once you are a miner, it is extremely difficult to find alternative work.

5.3.3 Waving the previous standard of living goodbye

A shortage of income had severe effects on the participants' living standards. We heard about the erosion of institutions and services that should provide an acceptable living environment. At Blyvooruitzicht mine, retrenchments affected the participants' basic living needs: food, water, housing, electricity, sanitation and maintenance services. One of them said the main challenge he faced was 'not knowing where the next plate of food would come from'.

Access to electricity, water, sanitation and the removal of rubbish became a problem in Blyvooruitzicht as the basic services previously provided by the mine owners, in exchange for a small monthly fee deducted from miners' pay cheques, were discontinued. A participant said the mining company used to pay for electricity. After the mine closed, the community tried to negotiate with the electricity provider, but it terminated the supply. The electricity situation was worsened by the activity of illegal miners, the *zama zamas*. In the Blyvooruitzicht mine and later in the village, they stole cable and underground pipes (excluding water pipes) and sold them for scrap. The mine's infrastructure was damaged, as can be seen by the many trenches in the area.

Water was also a problem. The municipality cut off the water and suspended sanitation, maintenance and waste collection services. Some participants tried to fix matters themselves. One participant said:

> In the Blyvoor community, the living conditions are not conducive to living in, because there are always sewage spills in the street due to sewage pipe blockages. We tried to fix it ourselves, but it was not easy sometimes. The municipality did not come to conduct essential maintenance.

The Lawyers for Human Rights referred two cases to the North Gauteng High Court in Pretoria, and the Blyvooruitzicht community won both. The Court instructed the electricity provider to install prepaid electricity metres for the households and the municipality to restore the water.

Retrenchment often has severe implications for a miner's housing situation, as mining communities often live in houses that belong to the mine. Most of the Blyvooruitzicht mine employees lived in company-owned hostels and houses in the mine village (LHR & FIDH, 2016). Seven of these participants said they had been evicted from their mine-owned houses after being retrenched and had nowhere else to go. Securing accommodation for their families became a priority. One participant said the community had 'panicked because they have no place to go'. Another was anxious because she did not have a title deed for the house. She said, 'Any time they can chase me away. Where will I go?'

However, some participants said they had refused to move out of the houses until they received their outstanding salaries and retrenchment packages. They felt they were entitled to live in the mine houses rent-free. One participant said they would stay 'because they owe us' and because there was nowhere to go. Another reason for being unwilling to leave was that they felt connected to the area.

For those who had bought their own houses, failure to make mortgage payments threatened their ability to provide shelter for their families. One of them spoke of 'a challenge with my bond account's repayment because I could not afford to pay'.

Participants in the Goldfields study who were homeowners tried to secure their houses by using their retrenchment package to pay part of their mortgage. Six participants had used the money to buy houses, paid part of their mortgage or bought sites and building materials. However, they did not have enough money to construct the planned houses. One said, 'we bought a site with the intention of building a house', but when the income disappeared, they were reduced to buying some zinc sheets to build shelter. Another said, 'the mines decided to sell the homes that we had been renting all along', and so they used the retrenchment money to buy the house.

One participant said she had bought the house they were renting from the mines with the retrenchment money, renovated it and was able to occupy it in December 2019. She told herself that even if the retrenchment money was all spent, at least she would know she had not wasted it, because, as she said, 'my kids would have a place they called home'.

The local municipality claimed in court proceedings that it could not assist the community in Blyvooruitzicht, as their dwellings were on land privately owned by the developer who had bought it from the mine, but this claim was overturned.

Social disruption (see Chapter 2) in the Blyvooruitzicht community had brought lawlessness and crime. Rape of women was mentioned specifically. Many said the area was unsafe. When the mine was still in operation, it provided private security that kept crime to a minimum. After the mine closed, the state police force was supposed to take responsibility for the safety of the area. But participants said the police refused to conduct night patrols as the area was too dark and unsafe, so the community took this upon themselves. One said that some had 'lost their lives because *zama zamas* shot them at night' and 'others were lucky to survive'. Another thought was that it was better to leave, 'because at night in the area it was not nice', and he decided he would go back to the labour-sending area he originally came from.

5.3.4 Health concerns

Several Blyvooruitzicht participants were concerned about their health. They believed that the mine did not care about their physical well-being while it employed them, and certainly not after it retrenched them. One participant who had contracted TB had been told to go home without any compensation for medical treatment and to come back when he was cured. When he returned, he discovered that 'the mine was retrenching people – until the manager left and then the mine was closed'.

Most of the Goldfields participants did not belong to a medical aid scheme while employed at the mine but had to use their wages for healthcare. One said that since the mine did not provide medical aid, he had been 'blessed' that his family had not needed medical treatment. After retrenchment, it was evident that they could not afford healthcare. Two participants said they had been diagnosed with underlying health problems post-retrenchment. One participant mentioned that they were being affected by silicosis. Anxiety about TB was common. Two participants who had been diagnosed with TB while still employed at the mine had to go for TB screening after their retrenchment. One TB sufferer who needed annual check-ups said: 'I am going through the screening treatment programme that includes better screening and treatment for miners who have been at risk of TB in the mines and those previously affected by this disease'. He said that when he was working, he had been able to afford adequate healthcare for his family and himself 'because I was earning a decent income'. He was worried that if his next check-up showed he had TB again, his family might get infected and would not 'have as much access to healthcare' and would be unable to 'follow treatment or get used to it'.

Two Goldfields participants said their spouses had died from ill health after their retrenchments. They felt this would not have happened if they had still been employed and able to afford the required medical treatment.

5.3.5 Psychological distress

The retrenched miners experienced shock, distrust, vulnerability, stress, disappointment, anger, hopelessness and depression (all examples of the effects of social disruption). Many of them were struggling to deal with these emotions without access to professional psychological support, which they could not afford. One participant said that being able to share their hurt and their problems with the interviewer brought a sense of relief, indicating the need for sharing and support. A Blyvooruitzicht participant said that 'to be affected by liquidation was very painful because you know you have a job, but the next day you wake up with no job'. Another said that people had died 'because they had experienced stress and other problems'. He said that if he had not been strong physically, he 'would have passed away too because of stress'. He said, 'I was strong mentally to deal with depression and other challenges'.

5.3.6 Damage to families

Several participants from both areas gave examples of family disintegration because of either marital problems or migration (divorce rates are often used as an indicator of social disruption). Some participants reported tension in the household. One participant worried that 'if you are a man and not working, your wife could disrespect you'. Another said:

> The retrenchment was very severe to my family, as I was new in a marriage. With no income and so much more time in those days, it left us poor and weak. Whenever it gets down to the purchasing of groceries, that is where you get the hard blow and are ashamed as the man and head of the family.

One said, 'Right now, my ex-wife is married somewhere, and I am staying alone', and another said the situation had affected him 'because I ended up separated from my wife'.

In contrast, one participant from Blyvooruitzicht said that although a man's partner will often leave or divorce him after he becomes unemployed, his situation was different. He was fortunate, as his wife had stayed and supported the family. Some women had chosen to stay with their husband or partner because they valued their families and trusted that things would get better.

Some participants said retrenchment had reinforced migration. Eleven of the 15 Goldfields participants had originally lived outside the Free State province. After being retrenched, eight had found work elsewhere and had had to leave their families behind. Retrenched miners' children are thus often faced with growing up without their mother or father. The disruption to families in the Goldfields was evident in comments like these:

> I am also missing raising my family as they are far away in the Eastern Cape. I cannot carry them because I am moving from one place to another, searching for work in the mines.

> Everyone understands that the cost of living has been incredibly high these days. Therefore, I was forced to leave my wife and children behind and went back to Welkom to search for work.

Paying school and university fees was a common concern. Social disruption studies often find poorer educational outcomes and early drop-out rates (see Chapter 2). Six participants from Blyvooruitzicht said they could not afford their children's school and tertiary education fees, so the children dropped out. Six Goldfields participants said they could not afford to send their children to a tertiary education institution for further studies, and three said they had to take their children out of the private schools that they attended and enrol them in government schools. In three instances, the miners' children had to leave their schools due to unpaid school fees. Two Goldfields participants elaborated. One said he had planned to pay for tuition for a student his grandmother was caring for but now could not do so because 'I am also struggling with my two children at the university'. The other said she was responsible for one of her siblings who had completed matric. She had told him he could not go to a tertiary education institution 'because I didn't have an income due to retrenchment'. One Goldfields participant said he had found odd jobs on farms and had withdrawn his children from school. Another said he had moved to a farm for work. Initially, he enrolled his children at a farm school, but later due to poverty, they were forced to leave the farm school at a young age in order to work on the farm as well.

Family outings had to be suspended for lack of money. Another Goldfields participant said, 'it was difficult for the kids because some of the things we used to buy for them must be limited or cut'. He said that 'this was the time when I felt poverty'.

5.3.7 Strategies for coping

The participants mentioned several coping strategies that had helped them survive. Several Goldfields participants received assistance from their spouses who had employment outside the mining industry. The spouses took over debt repayments and covered other expenses. One said his wife had to 'take all the responsibilities for my debts because she was the only one working', another said, as his wife was a teacher, 'she had to take over the mortgage bond', and another said, as his wife was still working, 'she now needs to pay for services in Virginia and the other house in Doorn'. Eight participants from the Goldfields area said their standard of living had deteriorated and they could no longer provide for their families as they used to but had to scale down and restrict their spending. Family support was important. Several participants said their families helped with food, school fees and debt repayment and provided money when needed. One said he used to assist his family with money, but 'now I had to request money from the family', and another said his sister helped him, 'mostly when I failed to pay for my children's school fees'.

Temporary or informal jobs ('piece work') helped some participants cope. Many Goldfields participants said they relied on these to survive. But such jobs were not easy to find. Most participants said that limited skills, a low education level or old age restricted their options, even though they were prepared to do any job. Some had had to accept low-paying, difficult or risky jobs. This temporary work occasionally provided for basic needs, as one participant explained:

> I survived through a piece job. I sometimes dig trenches. I get paid R15 [approximately $1.05] per metre. When I got R15, I could buy a meal to have food on the table and buy toiletries.

Some participants lived on income from renting out rooms in their house. One said he was renting out two bedrooms in a three-bedroom house. He said, 'the tenants pay R400 per room a month so that I can put food on the table'. Another said her mine-owned house had three rooms, one of which she divided, and 'at least I have got two tenants'. However, they did not regard this as a long-term solution as their tenants might move out, and they felt they could not complain when tenants behaved badly, as they depended on the rent for survival.

Some participants had to use all their savings to survive while unemployed. Many participants in both study areas depended on SASSA grants, including child support grants, to provide for basic needs for themselves and their families. These social grants are often the only source of reliable income for many South Africans from low-income families, especially in times of financial distress.

5.4 What can be done differently?

The findings of the two studies showed the far-reaching effects of mine downscaling and closure on employees, their families and the mining community. We wondered whether it is possible to minimise a mining community's dependency on the mine and manage the social disruption caused by mine closure differently. Lessons can be learned from other research (e.g. Ackermann et al., 2018; Coffie et al., 2021; Ndeleki, 2018), which highlights the role of government and top management and the importance of infrastructure investment after mine closure in mining communities. The role of the government is to develop awareness programmes that communicate the long- and short-term implications of retrenchment on individuals, families and communities. Owners of mining companies could obtain mine closure insurance and ensure retrenchment regulations in the Labour Act are followed, including negotiation processes with the input of various stakeholders. The unions can play a significant role in this process, and the miners should be involved in developing a realistic mine closure plan. Investment in infrastructure and social services (such as healthcare, education, service delivery and housing) in these communities after mine closure could curb social disruption. Various stakeholders (such as local government, the private sector and communities) could investigate alternative social and economic opportunities, such as diamond cutting and polishing, and, although limited, opportunities in the renewable energy sector. Retirement villages and holiday resorts could provide jobs, transform the mining-led economy and create a sustainable post-mine-closure economy. Depending on the geographical location of these communities, these alternatives could be viable replacements for mining for at least a portion of the retrenched miners.

Despite a strong call to increase miners' employability in other industries by helping them transfer their skills to similar industries or equipping them with the entrepreneurial skills needed for self-employment, these do not always prove to be feasible solutions. A Namibian study (Ndeleki, 2018) showed that training is not often available in remote areas. And even if former miners do learn additional skills, there are not necessarily job opportunities for them to pursue. Ndeleki (2018) emphasises the need to provide miners with some kind of proof of training, such as a certificate. However, even if former miners receive training, they do not always have the tools or equipment needed, and not all have the entrepreneurial aptitude to run their own business. Training alone is not necessarily the answer unless obstacles such as these receive attention.

5.5 Conclusion

Job losses due to mine downscaling and closures in developing countries like South Africa have severe implications. The loss of a mine that a town depended on as its sole economic driver leads to the termination of basic services such as electricity, water, sanitation and maintenance. Retrenched miners lose their houses and cannot pay for education and medical care. Our participants spoke of drastically reduced living standards, struggles to provide food for their families and threats to their safety. Many former miners are separated from their families because of family disintegration and migration. The evidence from our two case studies points to severe social disruption. Our participants mentioned increased crime and being enticed into getting involved in crime.

The long-term path dependency that a mining company can create and the social disruption that results when the company withdraws were evident from the interviews. In the case of Blyvooruitzicht, the sale of the mine to another company and the sale of its attached village to a developer had equally dire consequences for the community. The participants' responses showed that they and their families have had to align their way of life with their reduced circumstances and cope with an increase in crime. The government and the mining companies clearly need to do more to help miners deal with mine downscaling and closure. Current policy and the Draft Mine Closure Strategy (DMRE, 2021; see Chapter 3) do not consider all aspects of the problem.

Acknowledgement

This chapter draws substantially on empirical studies conducted by Boitumlo Joy Alec (2020) in the Free State Goldfields and Mmboneni Magadzu (2021) at the Blyvooruitzicht gold mine on the West Rand in the course of their master's studies at the Centre for Development Support, Faculty of Economic and Management Sciences, University of the Free State.

References

Ackermann, A., Botha, D. & Van der Waldt, G., 2018. Potential socio-economic consequences of mine closure. *The Journal for Transdisciplinary Research in Southern Africa*, 14(1), p. a458. https://doi.org/10.4102/td.v14i1.458

Alec, B., 2020. Mine closure narratives in the Free State Goldfields. Unpublished master's dissertation in Development Studies, University of the Free State, Bloemfontein.

CDS (Centre for Development Support), 2006. Towards a post-mining economy in a small town: Challenges, obstacles and lessons from South Africa. In *Centre for Development Support, Faculty of Economic and Management Sciences*. Bloemfontein: University of the Free State.

Chandra, D., 2014. A study on socioeconomic impact of retrenchment on workers – With special reference to Allahabad. *Management and Labour Studies*, 39(2), pp. 229–248.

Coffie, R.B., Sardiya, A., Ellis, F. & Akaadom, E., 2021. Can downsizing correlate with community development? A complex adaptive systems approach. *Inkanyiso, Journal of Human and Social Sciences*, 13(2), pp. 181–198.

Cornelissen, H., Watson, I., Adam, E. & Malefetse, T., 2019. Challenges and strategies of abandoned mine rehabilitation in South Africa: The case of asbestos mine rehabilitation. *Journal of Geochemical Exploration*, 205, art. no. 106354.

DMRE (Department of Mineral Resources and Energy), 2021. *Draft National Mine Closure Strategy*. Pretoria: DMRE. www.envass.co.za/draft-national-mine-closure-strategy-2021/

Haney, M. & Shkaratan, M. 2003. Mine closure and its impact on the community: Five years after mine closure in Romania, Russia, and Ukraine, Policy Research Working Paper 3083. World Bank, Europe and Central Asia Region: Infrastructure and Energy Services Department.

IPPR (Institute for Public Policy Research), 2017. Beneficiation in Namibia: Impacts, constraints and options. Windhoek, Namibia: IPPR.

LHR & FIDH (Lawyers for Human Rights & International Federation for Human Rights), 2016. *Blyvooruitzicht Mine Village: The Human Toll of State and Corporate Abdication of Responsibility in South Africa*. www.lhr.org.za/wp-content/uploads/2020/01/Blyvoor-HRIA_Final-Low-Def.pdf

Maeko, T., 2020a. Blyvooruitzicht: A glimmer of hope between the sewage, rats and fear. *Mail & Guardian*, 14 February.

Maeko, T., 2020b. Blyvoor Gold to resume operations in April. *Mail & Guardian*, 14 February.

Magadzu, M., 2021. Mine closure narratives in the West Rand. Unpublished master's dissertation in Development Studies, University of the Free State, Bloemfontein.

Marais, L., 2013. The impact of mine downscaling on the Free State. *Urban Forum*, 24, pp. 503–521.

Marais, L., 2023. *The Social Impacts of Mine Closure in South Africa: Housing Policy and Place Attachment*. London: Routledge.

Marais, L., Ndaguba, E., Mmbadi, E., Cloete, J. & Lenka, M., 2022. Mine closure, social disruption, and crime in South Africa. *The Geographical Journal*. doi: 10.1111/geoj.12430

Marais, L., Pelser, A., Botes, L., Redelinghuys, N. & Benseler, A., 2005. Public finances, service delivery, and mine closure in Koffiefontein (Free State, South Africa): From stepping stone to stumbling block. *Town and Regional Planning*, 48(S), pp. 5–16.

Martikainen, P., Mäki, N. & Jäntti, M., 2007. The effects of unemployment on mortality following workplace downsizing and workplace closure: A register-based follow-up study of Finnish men and women during economic boom and recession. *American Journal of Epidemiology*, 165(9), pp. 1070–1075.

Mörk, E., Sjögren, A. & Svalery, H., 2020. Consequences of parental job loss on the family environment and on human capital formation: Evidence from workplace closures. *Labour Economics*, 67, art. no. 101911.

Ndeleki, C.N., 2018. The effects of mine downscaling (and closure) on the socioeconomic development of mining communities: The case of Oranjemund. Unpublished mini-dissertation, MA (Development Studies), Centre for Development Support, Faculty of Economic and Management Sciences, University of the Free State.

Nichols, A., Mitchell, J. & Lindner, S. 2013. *Consequences of Long-Term Unemployment*. Washington, DC: The Urban Institute.

Pohlan, L., 2019. Unemployment and social exclusion. *Journal of Economic Behaviour and Organization*, 164, pp. 273–299.

Posel, D., Oyenubi, A. & Kollamparambil, U., 2021. Job loss and mental health during the COVID-19 lockdown: Evidence from South Africa. *PLoS One*, 16(3), art. no. e0249352.

Rixen, A. & Blangy, S., 2016. Life after Meadowbank: Exploring gold mine closure scenarios with the residents of Qamini'tuaq (Baker Lake), Nunavut. *The Extractive Industries and Society*, 3, pp. 297–312.

Tella, O. 2016. Understanding xenophobia in South Africa: The individual, the state and the international system. *Insight on Africa*, 8(2), pp. 142–158.

6 West Rand

Decline in South Africa's economic heartland

Margaret Kusambiza-Kiingi, Verna Nel and Job Gbadegesin

6.1 Introduction

Johannesburg has largely been successful in transitioning from a mining economy to a secondary and later tertiary economy (in the form of financial services). In the 1930s, when gold mining in Johannesburg shifted from the East Rand to the West Rand, the East Rand's economy shifted to the manufacturing sector. The West Rand, however, where the mines began to decline in the early 1990s, has not had the same success. It is struggling to deal with the consequences of the decline and find alternative economic options.

Gold mining used to be the backbone of the South African mining industry. The West Rand economy depended largely on it until decline set in. Factors causing the decline are the volatile gold prices, a deterioration in the quality of the gold deposits, and the increased cost of production, mainly because of deeper mining and labour unrest (Neingo & Tholana, 2016). The globalisation of the mining industry in South Africa since the 1990s has contributed to the decline (Crankshaw, 2002). Global shareholders emphasised the financial bottom line and limited spending on non-mining activities. In 1996, 55% of the West Rand economy depended on mining. By 2018, this had dropped to 30% (WRDM, 2021). Between 1996 and 2019, mining activity declined by 79%, causing the West Rand's overall economy to decline by 29%. This has had knock-on effects in the form of job losses and the closure of businesses. But it also affects socioeconomic stability.

This chapter describes the nature, scale and consequences of mine closures on the West Rand. We focus on the three local municipalities in the area: Merafong City, Mogale City and Rand West City. The inability to close mines appropriately has severe socioeconomic consequences. Policymakers chase dreams of economic transition. We argue that it would be better to accept economic decline as a reality that can be planned for.

6.2 The nature and scale of mining and mine closure

We identified 29 mines on the West Rand: 10 in Merafong City, 8 in Mogale City and 11 in Rand West City in 2018 (S&P, 2019) (Figure 6.1). Gold dominates the mining sector of the West Rand, constituting 72% of the total mines, followed by

DOI: 10.324/9781003403326-6

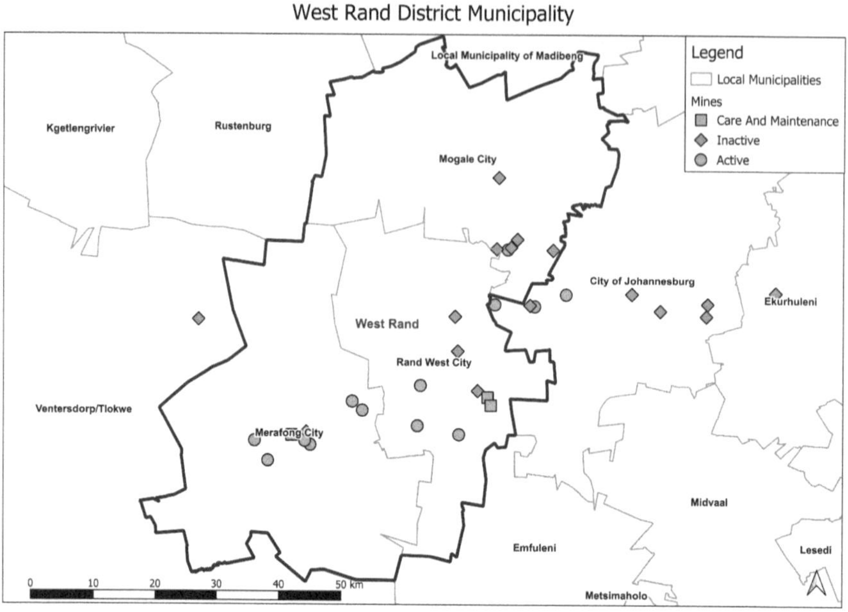

Figure 6.1 Mines and settlements in the West Rand

uranium (17%), diamonds, platinum and nickel. Yet 57.1% of the inactive mines are gold mines. Five companies have gold mines on the West Rand: Sibanye, Anglo Gold Ashanti, DRDGold, Harmony and Blyvoor Gold.

Mine closure, actual and imminent, comes in different forms; we identified four on the West Rand: active closure, care and maintenance, mining the residue, and selling to other firms. We found that only 12 (41%) of the original mines were still active. Fourteen (48%) were inactive (i.e. closed), eight of these being in Mogale City. Only 3 (11%) of the 29 were in care and maintenance, avoiding the cost of closure. When underground resources are no longer profitable, companies may mine the tailings and dumps. Seven of the 12 active mines were doing this. A rapid change in mine ownership indicates a declining industry (Vivoda et al., 2019). Large firms become unprofitable, selling to smaller firms whose cost structure is less expensive. Changing mine ownership is common on the Wet Rand. Between 2000 and 2019, one of the mines changed ownership six times. A further six changed ownership two or more times.

6.3 Consequences of mine decline and closure

Mine decline brings in its wake a complex set of interlinked demographic, economic, social, environmental, health and financial consequences. Towns that are losing their economic foundations find it difficult to accept the loss of their historical dependence on mining and to cope with the ensuing social disruption.

6.3.1 Demographic consequences

Mine decline on the West Rand has caused three kinds of demographic change. First, population growth has slowed, and in some cases, populations are decreasing. Between 2011 and 2016, the West Rand had an annual population growth of 0.43%. For comparison, in the same period, the population growth in the Sedibeng district, south of Johannesburg, was 0.88%, and the overall growth rate for the Gauteng province was 1.77% per annum (Stats SA, 2016). Merafong City had a negative population growth rate and an outflow of people; from 2011 to 2017, 23,545 people left and only 662 arrived (MCLM, 2021). Second, the dependence on migrant labour has declined. The 2016 data show that 91.6% of West Rand people born outside South Africa were from SADC countries, compared to 83.1% in Ekurhuleni (East Rand), 80.8% in the City of Johannesburg and 80% each in Sedibeng and the City of Tshwane (north of Johannesburg) (Stats SA, 2016). Net migration from Gauteng was 6.9% lower during 2016–2021 than during 2011–2016 (Gauteng Department: Treasury, 2022). Third, the high sex ratio (the ratio of males to females in a population) caused by an influx of men seeking jobs on the mines has changed. On the West Rand in 1996, the male-female ratio was 140:100. By 2016, owing to mine decline, it had dropped to 108:100.

A paradoxical effect of a decreasing population (and the ensuing economic downturn) is increased land use. Some reasons for this are the inability to afford formal housing, leading to sprawling informal settlements; new household formations being smaller in size; land invasions and land releases by municipalities.

6.3.2 Economic consequences

Mine closure has rapidly reduced the contribution the West Rand mines made to the local economy. In 2019, mining GVA was only 29% of what it was in 1996, and the overall economy was only 70% of what it was in 1996. Mine employment plummeted: Global Insight (2021) estimates a decline from 110,000 in 1996 to 40,000 in 2019, and overall employment declined by 14% annually. SARS tax data show a decrease in mine employment from 80,000 in 2014 to 62,000 in 2021, but a decrease of only about 6,000 in overall employment (SARS, 2023). There are differences between the three municipalities. Merafong suffered most, losing about 11,000 mining jobs and 11,000 jobs overall. Mogale City saw employment increase from 85,000 to 96,000, despite the loss of 2,000 mining jobs, and it managed to create about 6,000 new jobs in manufacturing. Rand West City had a small increase in mining jobs but a decrease of 6,000 jobs overall. Consequently, the West Rand has experienced an increase in unemployment. Global Insight (2021) estimates an increase in unemployed people from 48,000 in 1996 to nearly 139,000 in 2019. Poverty levels have increased, and there has been a shift to informal work. The Gauteng provincial treasury reported declining rates of new investments in the West Rand, with a 4% decline in the growth rate of investments from 2016 to 2020 (Gauteng Department: Treasury, 2022). This drop has been attributed to the decline in mining.

6.3.3 Social consequences

One of the premises of this book is that mine closure brings social disruption. This disruption is due partly to mines closing and partly to a failure to regulate mine closure.

6.3.3.1 Social unrest

Social protest is common in South Africa. Protests are very common on the West Rand. Lack of data makes it impossible to tell whether it has more protests than non-mining areas. The University of Johannesburg's Centre for Social Change found a relationship between a region's dependence on natural resources and violence (Runciman et al., 2016). Violent protests on the West Rand result from poor service delivery and an inability to access land. Media reports mention the torching of buildings (including a clinic), protests about the lack of water, clashes with the police and riots over land in Rietvallei in Mogale City (ENCA, 2013; Zulu, 2017; Seleka, 2021). Informal settlements are hotspots for social unrest linked to inadequate service delivery, unfulfilled promises by politicians and mining companies, mistrustful relationships between role-players, and long-standing complaints (Visagie et al., 2021). Van Eeden and Khaba (2016) found that protests have a political angle. It has been claimed that politicians exploit xenophobia to instigate protests.

6.3.3.2 Land invasions and informal settlements

Increased unemployment, poverty and income insecurity have forced more people to live in informal settlements (Cole & Broadhurst, 2022). Urban sprawl is a problem in both growing and declining mining areas (Marais et al., 2020). Informal settlements mushroom around active and inactive mines, posing a challenge for local municipalities. On the West Rand, the land occupied by informal settlements has more than doubled, from 476 hectares in 1990 to 1,257 hectares in 2020. Some settlements are on mine land. In January 2014, for example, more than 600 residents from Bekkersdal (Mogale City) occupied vacant land belonging to Rand Uranium (Mathebula, 2014). The percentage of households living in informal housing remains high: 22.5% on the West Rand compared with 18% for the whole of Gauteng province (Stats SA, 2016). The National Department of Human Settlements identified 116 informal settlements on the West Rand, compared to 33 in Sedibeng, 115 in Tshwane and 123 in Ekurhuleni (Parliamentary Monitoring Group, 2017). The reasons for the development of informal settlements in mining areas are complex, with the growing informal settlements in mining areas that are in decline being more puzzling. Initially, the closing of single-sex mine hostels contributed, but there is evidence of people being unwilling to move away from the West Rand.

6.3.3.3 Crime and illegal mining

Marais et al. (2022) have shown a relationship between mine closure and crime, finding much higher crime levels in areas of mining decline than in areas of mining

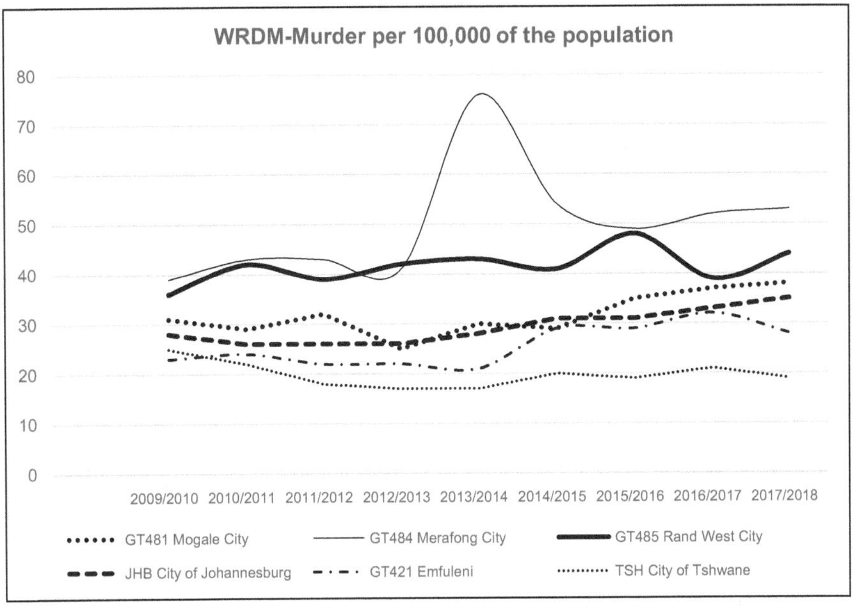

Figure 6.2 Murder rates on the West Rand compared with three other municipalities in Gauteng, 2009–2019

Source: Marais et al. (2022); crime data provided by the South African Police Service

growth. Figure 6.2 shows the murder rates per 100,000 of the West Rand population in comparison with other municipalities in Gauteng.

The murder rates are particularly high in Merafong City and Rand West City, substantially higher than those in Johannesburg. We highlight five reasons. First, the post-apartheid policy emphasising housing stability contributes (Marais, 2023). Mineworkers affected by closure usually fail to find alternative employment and cannot afford to remain in their housing. But when the local municipality makes land available, they take that opportunity in the hope of eventually finding a job. The result is a growing population despite mine closure, and generally, growing populations fuel crime. Second, policies focusing exclusively on mining growth and ignoring closure make underequipped and cash-strapped municipalities responsible for preparing for decline and dealing with each new problem. Third, even when local strategic plans consider mine closure, the emphasis is on environmental impacts. After closure, local planners fail to integrate the social aspects. Fourthly, unregulated closure encourages illegal mining activities, usually run by criminal syndicates (Marais et al., 2022). Illegal mining is a big problem for the West Rand, contributing to the high murder rate. Syndicate members often kill one another to protect their mining areas. Finally, poor service delivery contributes to rising crime. Badly maintained roads and the lack of house numbers make policing difficult. Lack of streetlighting makes it unsafe to walk at night, especially for women. Gender-based violence is common in mining communities and worsens with mine closure.

The media often reports on efforts to reduce crime. This includes partnerships between the police, the mines and private security companies to arrest criminals and seize illegal mining equipment (Creamer, 2019; *The Citizen*, 2019; Pretorius, 2021).[1] For example, in July 2022, the police raided Westonaria, arresting 800 suspects.[2] In August 2022, the police arrested about 100 *zama zamas* (illegal miners) suspected of the rape of eight women near an unused mine dump in Mogale City (Rademeyer, 2022). Yet affected communities blame the police for the high crime levels (Gumede, 2022).

6.3.4 Environmental and health consequences

Despite the economic value of gold mining, mine closure has had devastating environmental and human health consequences for the West Rand (Lusilao-Makiese et al., 2013). Settlements near tailing dams have exposed people to radon exhalation, radiation, dust and floods (Winde & Stoch, 2010). Living among these dumps is an example of material dependency. Abandoned mines have increased sinkhole hazards (Winde & Stoch, 2010). Kritzinger (2017) noted that 50% of the 2,500 sinkholes in South Africa are in the West Rand. Historical underground mining is the main contributing reason.

Although mining companies must comply with South Africa's mine closure legislation, several problems exist. Most closed mines operated under pre-2002 legislation, which had minimal formal mine closure and land rehabilitation regulations. Companies abandoned these mines. But mine closure for those under post-2002 legislation is difficult. Some mines opt for care and maintenance, while the government does not facilitate proper closure in other cases. Despite the disputes about closing mines, communities near mine tailings have to live with the consequences (AGSA, 2009).

The Parliamentary Portfolio Committee investigated a case of pollution in Mogale City by Mintails South Africa (mining old tailing dams and dumps) (Olalde & Matikinca, 2018). The Committee found that Mintails had failed to comply with the conditions of the mining permit and ignored its commitments in its social and labour plans. The rehabilitation cost was estimated at R383 million (Olalde, 2018). When the company was placed under business rescue and eventually liquidated, the hope for environmental rehabilitation vanished. Liquidation has become a way to avoid long-term rehabilitation requirements.

6.3.5 Financial consequences

Municipalities suffer financially when a mine closes. The West Rand municipalities became overly dependent on direct or indirect mine revenues. Mine closures reduced their tax income, with the result that they increasingly rely on transfers from the national government (Zwane, 2021). Figure 6.3 shows property rates collected in the three municipalities.

By 2019, two municipalities (Merafong and Rand West) had seen a decline in property rate collections. The decline is associated with people losing their jobs

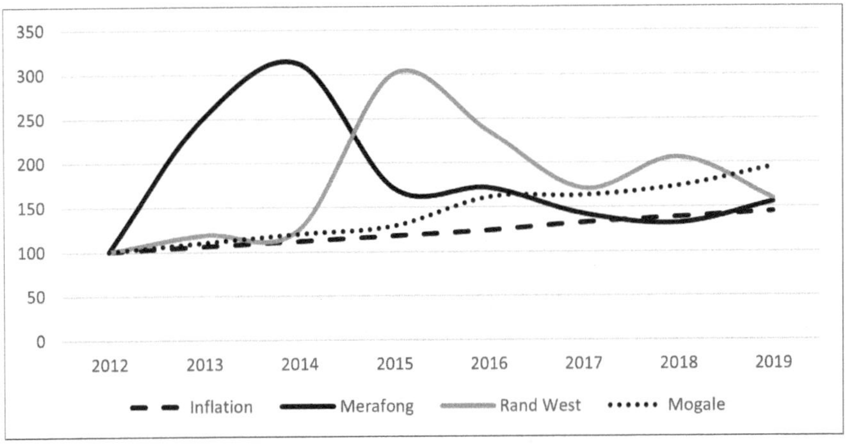

Figure 6.3 Property rates collected per local municipality compared to national inflation, 2012–2019 (2012 figures expressed as 100%)

Source: Marais et al. (2022); crime data provided by the South African Police Service

Table 6.1 Aged creditors and irregular expenditures, average for 2017–2019

	Merafong City	*Rand West City*	*Mogale City*
Irregular expenditure as % of total expenditure	15.8	11.1	5.8
Aged creditors as % of total expenditure	46.2	46.6	20.2
Debtors as % of income	68.3	35.2	53.9

Source: Data derived from municipal financial statements

and subsequent business closures. These two municipalities have experienced volatile property incomes. However, Mogale's property revenue has risen consistently at a rate higher than inflation.

The municipal electricity rates show that Merafong and Rand West could not access electricity income at a rate higher than inflation for most of the period 2012–2019. In Merafong, electricity income has remained stagnant since 2012. This could be largely due to illegal connections set up by people unable to pay.

There are irregularities in municipal financial management. Irregular expenditure as a percentage of total expenditure was the highest in Merafong (16%). Aged creditors (i.e. unpaid bills) in Merafong and Rand West City were 46%, and debtors as a percentage of income were 68% in Merafong (see Table 6.1).

Mine closures affect the local government's finances and ability to provide services. Aside from a declining tax base, many residents simply cannot pay for services. The result is a municipality that does not have money to spend on economic alternatives to mining.

6.4 Conclusion

The West Rand has experienced a rapid decline in mining over the past three decades. This has led to a smaller population and negative economic growth. Unemployment has skyrocketed. There is ample evidence of social disruption associated with mine closure: violent protests, inability to provide services, land invasions, deteriorating municipal finance and high crime levels. There is a high likelihood of environmental health concerns increasing in the future. Urban planners and policymakers tend to favour strategies for growth and expansion. They do not acknowledge the reality of decline. The government's pursuit of economic growth despite the evidence of decline is an example of goal dependency. Ignoring the inevitable, its overall strategy continues to be economic diversification and finding new growth paths. There is an urgent need to acknowledge the reality of decline and find appropriate ways of dealing with it.

Many social and environmental concerns result from the inability to close mines appropriately. Appropriate mine closures and rehabilitation could make land available for new economic opportunities. Living among the mine dumps (a material dependency) has health and safety risks. It inhibits land use change that could bring economic development and help curb illegal mining.

Notes

1 **Nkosazana Dambuza for ENCA** (eNews Channel Africa) reported on this on 7 October 2021, 'Crime in SA: Illegal mining on the rise'. The video vividly illustrates the desperate struggles of the illegal miners. www.enca.com/news/crime-sa-watch-illegal-mining-rise.

2 This was reported by ENCA (eNews Channel Africa) in a video on 14 July 2022, 'Operation O Kae Molao: Police descend on West Rand'. www.enca.com/news/operation-o-kae-molao-police-descend-west-rand

References

AGSA (Auditor-General of South Africa), 2009. *Report to Parliament on a Performance Audit of the Rehabilitation of Abandoned Mines at the Department of Minerals and Energy*. Cape Town: AGSA.

The Citizen, 2019. West Rand to tackle illegal mining through mining forum. *The Citizen*, 24 July.

Cole, M.J. & Broadhurst, J.L., 2022. Sustainable development in mining communities: The case of South Africa's West Wits Goldfield. *Frontiers for Sustainable Cities*, 4, pp. 1–5.

Crankshaw, P., 2002. Mining and minerals. In Lemon, A. & Rogerson, C. (eds.), *Geography and Economy in South Africa and its Neighbours*. Burlington: Ashgate, pp. 63–80.

Creamer, M., 2019. West rand police did well to lift the lid off illegal mining "workshops", now Barberton violence perpetrators must be exposed. *Mining Weekly*, 19 July. www.

miningweekly.com/article/west-rand-police-did-well-to-lift-lid-off-illegal-mining-workshops-now-barberton-violence-perpetrators-must-be-exposed-2019–07–19.

ENCA (eNews Channel Africa), 2013. Fochville on edge after protests, *ENCA*, 6 August. www.enca.com/south-africa/fochville-edge-after-protests

Gauteng Department: Treasury, 2022. *Gauteng Province: Socio-Economic Review and Outlook 2022*. Johannesburg: Gauteng Provincial Government.

Global Insight, 2021. *Rex Explorer Database*. Pretoria: Global Insight.

Gumede, B., 2022. Tembisa protests, Krugersdorp anger raise fears of new wave of violence. *Mail & Guardian*, 2 August. https://mg.co.za/news/2022-08-02-tembisa-protests-krugersdorp-anger-raise-fears/

Kritzinger, A.S., 2017. Socio-economic impact assessment of the proposed cessation of pumping and associated closure of the underground workings of the Ezulwini operations of Sibanye Gold. In *Gauteng Province*, Westonaria: Sibanye Gold Ltd.

Lusilao-Makiese, J., Cukrowska, E., Tessier, E. & Amouroux, D., 2013. The impact of post gold mining on mercury pollution in the West Rand region, Gauteng, South Africa. *Journal of Geochemical Exploration*, 134, pp. 111–119.

Marais, L., 2023. *The Social Impacts of Mine Closure in South Africa: Housing Policy and Place Attachment*. London: Routledge.

Marais, L., Denoon-Stevens, S. & Cloete, J., 2020. Mining towns and urban sprawl in South Africa. *Land Use Policy*, 98, p. 103953.

Marais, L., Ndaguba, E., Mmbadi, E. & Cloete, J., 2022. Mine closure, social disruption, and crime in South Africa. *The Geographical Journal*, 188(3), pp. 383–400.

Mathebula, N., 2014. Bekkersdal land invasion: Community won't budge. *Randfontein Herald*, 31 January. https://randfonteinherald.co.za/149597/bekkersdal-land-invasion-community-wont-budge/

MCLM (Merafong City Local Municipality), 2021. *Merafong city local municipality review of 2016–2021 integrated development Plan*. Carletonville: Merafong City Local Municipality.

Neingo, P. & Tholana, T., 2016. Trends in productivity in the South African gold mining industry. *Journal of the Southern African Institute of Mining & Metallurgy*, 116(3), pp. 283–290.

Olalde, M., 2018. Mintails collapse: Case study in how not to close mines. *Oxpeckers Investigative Environmental Journalism*, 8 August. https://oxpeckers.org/2018/08/mintails-collapse/

Olalde, M. & Matikinca, A., 2018. Directors targeted for Mintails mess. *Oxpeckers Investigative Environmental Journalism*, 12 December. https://oxpeckers.org/2018/12/mintails-directors-targeted/

Parliamentary Monitoring Group, 2017. *List of 2453 Informal Settlements Gathered by NDHS [National Department of Human Settlements]*. https://static.pmg.org.za/RNW86-2019-07-16_Annex_A.pdf

Pretorius, N., 2021. Thirteen arrested in major illegal mining bust in West Rand District, *Randfontein Herald*, 14 January. https://randfonteinherald.co.za/371795/thirteen-arrested-in-major-illegal-mining-bust-in-west-rand-district/

Rademeyer, J., 2022. Zama zamas just one part of a multibillion-rand organised crime economy that threatens SA. *Daily Maverick*, 26 May. www.dailymaverick.co.za/article/2022-08-07-zama-zamas-just-one-part-of-a-multibillion-rand-organised-crime-economy-that-threatens-sa/

Runciman, C., Alexander, P., Rampedi, M. & Moloto, B., 2016. Counting police-recorded protests: Based on South African Police service data. *Centre for Social Change*. Johannesburg: University of Johannesburg.

SARS (South African Revenue Service), 2023. SARS tax data. Pretoria: SARS.

Seleka, N., 2021. "We don't want RDP houses, we want land": Crowd clashes with cops over Rietvallei land invasion. *News24*, 24 May.

S&P Global Market Intelligence, 2019. Mining database, S&P.

Stats SA (Statistics South Africa), 2016. *Provincial Profile: Gauteng Community Survey*. Pretoria: Stats SA.

Van Eeden, E.S. & Khaba, B., 2016. Politicising service delivery in South Africa: A reflection on the history, reality and fiction of Bekkersdal, 1949–2015. *Journal for Contemporary History*, 41(2), pp. 120–143.

Visagie, J., Turok, I. & Swartz, S., 2021. What lies behind social unrest in South Africa, and what might be done about it. *The Conversation*, 18 August. https://theconversation.com/what-lies-behind-social-unrest-in-south-africa-and-what-might-be-done-about-it-166130

Vivoda, V., Kemp, D. & Owen, J., 2019. Regulating the social aspects of mine closure in three Australian states. *Journal of Energy & Natural Resources Law*, 37(4), pp. 1–22.

Winde, F. & Stoch, E., 2010. Threats and opportunities for post-closure development in dolomitic gold mining areas of the West Rand and Far West Rand (South Africa): A hydraulic view: Part 1: Mining legacy and future threats. *Water SA*, 36, pp. 69–74.

WRDM (West Rand District Municipality), 2021. *West Rand District Municipality: Profile & Analysis: District Development Model*. Randfontein: West Rand District Municipality.

Zulu, S., 2017. Zenzele Informal Settlement residents damage 10 cars amid protest. *Eye Witness News*, 17 March. https://ewn.co.za/2017/03/17/zenzele-informal-settlement-residents-block-r28-with-stones

Zwane, T., 2021. Municipal finances in state of crisis. *City Press*, 31 October. www.news24.com/citypress/business/municipal-finances-in-state-of-crisis-20211031

7 Matjhabeng

Decline in the urban periphery

Kentse Sesele and Margaret Kusambiza-Kiingi

7.1 Introduction

Most mining companies now include decline and closure when planning a mine's life cycle. This is supposed to help mining cities anticipate problems and plan to deal with them. The solution that is usually suggested is to diversify the city's economy. This is difficult in the remote areas where mines are often situated. The few successes have been mainly in Europe or in countries where long-term state subsidisation buffers the impact. We question whether economic diversification is a realistic solution for remote mining areas in South Africa.

Many mining cities are facing the need to manage the economic consequences of mine closures (Martinez-Fernandez et al., 2012; Cortinovis et al., 2017). Resource depletion or changing market conditions can force a mine to close, bringing decline and decay to the city that depends on it, with 'contractions in production and investment, the relocation of capital, increased unemployment and population loss' (Li et al., 2015, p. 311). Many frameworks have been proposed for understanding economic change related to mining and mine closure. The research has examined economic and institutional change (North, 1990), the role of relatedness in finding new growth trajectories (Boschma, 2017), the effect of the global production network (GPN) on a country's spatial development (Atienza et al., 2021), and the role of institutions and local agents in creating economic change (Binz et al., 2016). Recent research has linked the literature on mine decline and shrinking cities (Martinez-Fernandez et al., 2012; Li et al., 2015), both of which emphasise the importance of local planning systems.

South Africa's once-thriving gold mining industry has experienced a rapid decline over the past three decades. In the mid-1980s, it employed about 560,000 people; today, it has shrunk to about 110,000. Many places that benefited from gold mining in the 1970s and 1980s have experienced a rapid decline since then (Binns & Nel, 2001; Nel & Binns, 2002; Marais, 2013a; Marais et al., 2017; Marais & De Lange, 2021). Four municipalities that depended heavily on gold mining, Matlosana, Matjhabeng, Merafong City and Mogale City, have gone downhill. None of these had diversified their economies away from gold mining. Our case study, the

DOI: 10.324/9781003403326-7

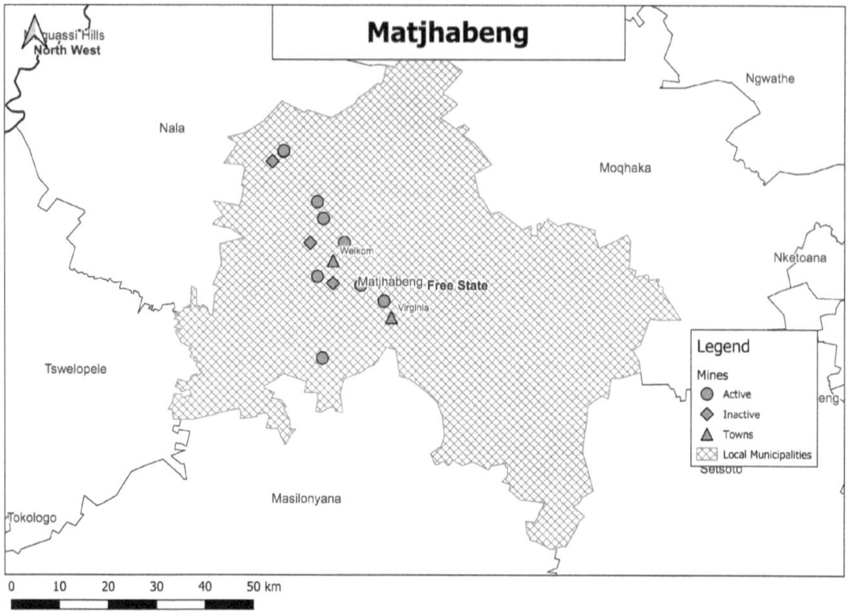

Figure 7.1 Mines and settlements in Matjhabeng

Free State Goldfields (today known as Matjhabeng), has had various plans to do so but has achieved little in this direction (Marais, 2013b; Marais & Nel, 2016) (see Figure 7.1). This chapter critically assesses the various attempts over the past six decades, with a detailed analysis of the latest economic development strategy. We argue that a combination of factors is at play: the lack of industries related to gold mining, the peripheral nature of the area and planning systems that inhibit progress. All these have contributed to the current state of affairs and blinded the planners to the value of the area's regional services function and the potential advantages of decline.

7.2 Methods

We analysed a new economic development plan for the Matjhabeng Local Municipality. In addition to this document analysis, we interviewed 15 key government, business and civil society representatives recruited through a snowballing process. We used thematic analysis to analyse the transcripts of the data. In addition, to profile the decline, we used economic data from Global Insight (2020) and historical employment data. Our study site was the area historically known as the Free State Goldfields. But data were sometimes available only for the Matjhabeng Local Municipality, while historical data were available for the Free State Goldfields (a slightly larger area, which includes parts of Masilonyana). We make these distinctions clear in the chapter.

7.3 The history of mining and mine decline

7.3.1 Early history

Large-scale gold mining began in South Africa in 1886 when prospectors discovered gold on the Witwatersrand. This laid the foundation for Johannesburg to become the economic heart of South Africa. The Johannesburg Stock Exchange, the country's banking system and the University of the Witwatersrand were direct spinoffs from the mining industry (Davenport, 2013). The discovery of gold on the West Rand in the early 1900s (extending to the southern area of what is today known as the North West Province) marked the beginning of a mining boom. When its mining later declined in the early 1900s, Johannesburg became South Africa's gateway city, hosting mining technology capabilities, financial services and company head offices.

With the discovery of gold in the Free State Goldfields after the Second World War, several new towns, such as Welkom, Virginia and Allanridge, sprang up. By the mid-1980s, the Free State Goldfields employed 180,000 people in mining and produced about a quarter of the gold in the 'free world' (Marais, 2013a). The first decline was experienced in the early 1970s. When the United States left the gold standard in the mid-1970s, the price of gold soared. Marginal mines became profitable overnight, and many more shafts were sunk (Marais & Nel, 2016). The second period of decline set in at the beginning of the 1990s with the slump in the international gold price.

7.3.1.1 The nature and scale of economic change and disruption

Table 7.1 shows the large-scale job losses (about 160,000) that have occurred in most economic sectors since 1987. However, jobs in the finance sector have grown, and although the manufacturing sector lost jobs between 1987 and 1996, employment in this sector has stabilised since then. Figure 7.2 compares Matjhabeng's Gross Value Add (GVA) from 1996 to 2019 with the GVA of the Free State Province and South Africa. Matjhabeng has experienced substantial economic decline. The first large-scale mine downscaling took place in 1991. However, despite the rapid decline, mining still contributes about 45% of GVA (down from 62% in 1996). Mining employment dropped from a peak of 61% of total employment in 1996 to 25% in 2019. The population has declined from about 480,000 in 1996 to about 430,000 today.

7.3.1.2 Historical attempts at economic diversification

Marais (2013b) and Marais and Nel (2016) provide a detailed overview of historical attempts to diversify the economy. Here, we highlight the most important aspects. As early as the 1960s, there was an understanding that economic diversification was crucial. A planning study concluded that 'unless diversification occurs and other functions are added, possibly in response to the markets that have been created, these towns are likely to stagnate and die when mining operations cease'

Table 7.1 Employment per sector in the Free State Goldfields, 1980–2019

Economic sector	*1980*		*1987*		*1996*		*2001*		*2011*		*2019*	
	n	%	*n*	%	*n*	%	*n*	%	*n*	%	*n*	%
Agriculture: FSGF	10 735	4.7	11 194	4.1	13 775	8.6	17 593	14.1	8 264	8.6	7 488	9.0
Agriculture: Matjhabeng					9 377	6.5	13 084	11.8	6 020	6.5	5 436	7.0
Mining: FSGF	154 343	67.8	184 634	67.1	97 374	61.0	43 144	34.6	26 069	27.0	20 369	24.4
Mining: Matjhabeng					89 130	62.1	38 979	35.1	25 157	27.2	19 576	25.4
Manufacturing: FSGF	7 679	3.4	10 105	3.7	5 896	3.7	7 497	6.0	6 168	6.4	6 031	7.2
Manufacturing: Matjhabeng					5 570	3.9	7 928	7.1	5 525	0.6	5 560	7.2
Electricity: FSGF	501	0.2	743	0.3	531	0.3	419	0.3	543	0.6	1 107	1.3
Electricity: Matjhabeng					492	0.3	386	0.3	479	0.5	955	1.2
Construction: FSGF	10 913	4.8	15 875	5.8	4 256	2.7	3 701	3.0	4 782	5.0	4 041	4.8
Construction: Matjhabeng					3 837	2.7	3 259	2.9	4 325	4.7	3 653	4.7
Trade: FSGF	14 107	6.2	17 587	6.4	11 667	7.3	19 028	15.2	15 699	16.3	13 911	16.6
Trade: Matjhabeng					11 035	7.7	17 947	16.2	14 011	15.1	12 412	16.1
Transport: FSGF	2 562	1.1	2 804	1.0	2 301	1.4	2 637	2.1	3 018	3.1	2 459	2.9
Transport: Matjhabeng					2 122	1.5	1 081	1.0	2 800	3.0	2 277	3.0
Finance: FSGF	2 007	0.9	2 519	0.9	9 106	5.7	9 979	8.0	11 370	11.8	8 830	10.6
Finance: Matjhabeng					8 565	6.0	9 242	8.3	20 463	22.1	9 600	12.4
Services: FSGF	24 661	10.8	29 717	10.8	14 796	9.3	20 839	17	20 491	21.3	19 402	23.2
Services: Matjhabeng					13 407	9.3	19 217	17	18 741	20.3	17 692	22.9
Total: FSGF	227 508	100	275 178	100	159 702	100	124 837	100	96 404	100	83 638	100
Total: Matjhabeng					143 535	100	111 123	100	92 521	100	77 161	100

Source: Marais (2013a) and Global Insight (2020)

Note: The data exclude employment at households. Historical data were available for the Free State Goldfields for 1980 and 1987. For Matjhabeng, historical data were available only from 1996. To calculate the Free State Goldfields data for 1996, 2001, 2011 and 2018, we combined data for Matjhabeng and Masilonoyana (Marais, 2013a).

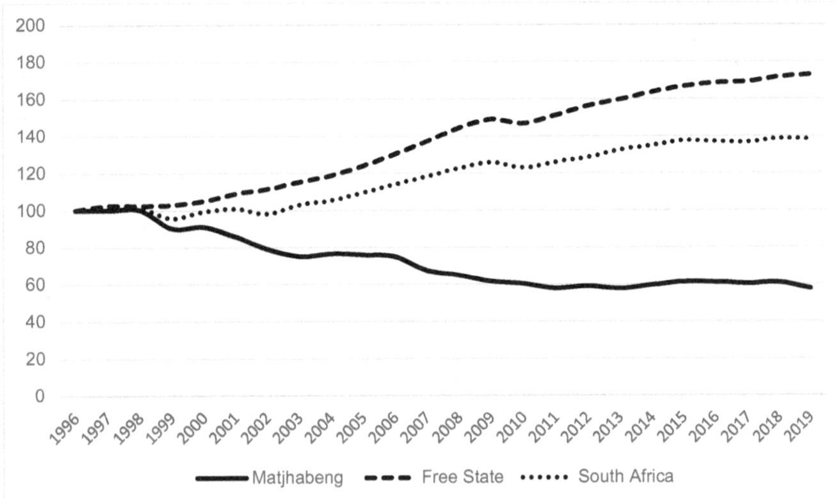

Figure 7.2 Changing GVA for Matjhabeng, the Free State Province and South Africa

Source: Global Insight (2020)

Note: We used the GVA for 1996 as 100% in compiling this and then recorded the change against the baseline.

(Page, 1969, p. 6). Consequently, the first economic development strategy appeared at the end of the 1960s. This initial strategy focused on four sectors: mining, industry, commerce and agriculture (Welkom Municipality, 1968). The prioritisation of commerce and agriculture laid the foundation for the development of a regional services function. However, when the gold price rose in the mid-1970s (when the United States left the gold standard), the municipality shelved the plan. Economic diversification took a back seat as mining wealth dominated.

The drop in the gold price at the end of the 1980s prompted a renewed attempt to think about economic diversification. A new structure plan highlighted the potential danger of mine closure (Welkom Municipality, 1989). Despite the likelihood of decline, however, the plan envisaged growth in the area to the south of Welkom and planned for a city of one million people. The plan assumed that more family housing would become available to mineworkers who historically resided in single-sex compounds. The structure plan recommended that the municipality develop an economic diversification strategy, focusing on agricultural industries, motor vehicle components and jewellery. A second recommendation was that the municipality should create a development agency outside of the municipality's control. Consequently, the municipality established the non-profit company, the Free State Goldfields Development Centre (FSGDC), with directors from the private and public sectors (Marais, 2013b). The focus was on the broader Free State Goldfields region, which included Welkom, Virginia, Odendaalsrus, Hennenman, Theunissen and Allanridge. The FSGDC has had some successes. For example, it helped create a local manufacturing industry with value chains outside of mining (mostly with the surrounding agricultural industry). The development of the Phakisa motor racing track, which hosted five MotoGP international motorcycle races between 1998 and 2003, was one of the most important success stories, but several challenges remained (Nel & Binns, 2002).

A more inclusive strategic plan was developed in 1998, after the democratisation of South Africa in 1994. Nevertheless, progress was hampered by a declining involvement by the mines, rapid changes in mine ownership and institutional instability in the municipal council (Marais, 2013b). A change in the legislation governing the relationship between government and municipal entities saw the FSGDC dissolved in 2004. Matjhabeng established a new entity in compliance with the legislation: the Matjhabeng Marketing and Investment Company (MMIC). The economic development function returned as a core municipal function. But conflict arose between the municipal economic department and the MMIC, and there was a call to establish a development agency in the Lejweleputswa district rather than in Matjhabeng. Consequently, the Lejweleputswa District Agency was established in 2005. The establishment of a district agency meant that the MMIC closed.

Overall, progress in economic diversification has been slow, with many projects not going beyond the planning stage. Yet the need for economic diversification remains. In 2014, the Matjhabeng mayor said, 'We are trying to move our focus away from mining to agriculture. We are also planning to boost tourism and employment in the green economy' (Bolowana, 2013). There were also attempts from the mining industry to help with economic diversification. In 2015, Harmony Gold contracted consultants, but this plan was never released.

Planning for economic diversification has been going on for nearly 60 years. There was an early start in the 1960s. The most important progress came in the late 1980s with the establishment of the FSGDC. The FSGDC made progress, but since its closure, the economic diversification process has been slow. Using an organisation outside of the municipal ambit is an example of the influence of the new public management paradigm. The functioning of the FSGDC resembles network governance, but this approach has not been pursued since the closure of the FSGDC. Overall, there is a tendency to focus on plan-making instead of implementing.

7.4 A new economic development strategy, 2019

A new economic development strategy saw the light in 2019. Harmony Gold paid for this strategy, and consultants developed it in cooperation with the Matjhabeng Municipality. The strategy's vision states that Matjhabeng 'aims to develop a globally competitive economy through the collaborative diversification of the mining, manufacturing, trade and finance sectors'. The strategy has five pillars: beneficiation, SMME (small, medium and micro enterprise) development, innovation and R&D (research and development), an enabling business environment and tourism. Each pillar has a set of objectives and related projects (see Table 7.2). The strategy pays little direct attention to mine closure and does not consider why historical strategies had limited success.

Table 7.2 Summary framework of the new local economic development strategy for Matjhabeng Local Municipality, 2019

Pillar	*Objective*	*Projects*
Beneficiation	Agro-processing	Aquaculture
		Poultry value chain development
	Diversify mining resources	Jewellery manufacturing
		Recycling plant
		Steel manufacturing
		Petroleum refinery
	Manufacturing incubation hub	Specialised economic zone
SMME development	Formalise the informal sector	Informal business complex
	Business intelligence	Links to funding support
		Business plan development
		Economic and small business development
		Establish business support centres
Innovation and R&D	Science and technology	Science park
		Science bursaries
		ICT infrastructure
		Digital economy
	International markets	Consulting Africa
		Cargo airport
		'Buy South Africa' campaign
Creating an enabling business environment	Improve ease of doing business	Simplify business regulation
		Create accessible local business database
	Training and skills development	Skills incubation hub
		Skills audit
		Job centre/skills bank
	Good governance	Develop sector plans
		Establish economic project investment committee
		Land parcel audit
Tourism	Tourism development	Sand river route
		Film incubation hub
	Tourism marketing	Travel guide
		Phakisa Raceway

Source: SACN (2020)

The first pillar, beneficiation (the transformation of a mineral into a higher-value product for local sales or export), has been a prominent strategy since the 1960s. Although progress has been made with a jewellery school, large-scale gold beneficiation remains elusive. The main reason is that the jewellery industry in South Africa is small and clustered in Johannesburg (Da Silva, 1999). Competing with the existing cluster is difficult, and businesses in the existing cluster are unlikely to relocate to Matjhabeng. In the absence of any other possible gold-based industry, the strategy proposes various other projects: aquaculture (linked to food processing), a poultry value chain (linked to food processing), a recycling plant, steel manufacturing, a petroleum refinery and the creation of a specialised economic zone in which manufacturing would be central. The recycling plant and the poultry value chain probably have the best chances of success as they link to local value chains. However, several concerns remain. Aquaculture is new in the area and not related to local industries. The steel industry in South Africa currently operates at low capacity levels, and plants are located either at the coast or where coal is available. Petroleum refineries in South Africa are largely in coastal cities. Furthermore, the creation of a specialised economic zone has been part of many previous strategies without success. The current reality is that the national government has allocated one specialised economic zone to the Free State, and the Free State Provincial Government has given it to Maluti-a-Phofung in the northeast Free State.

The second pillar is SMME development, which is a national priority. Although an SMME focus is essential to the diversification of the economy, there are two concerns. The strategy does not evaluate the current role of the small enterprise development agency (SEDA) and is vague on how the municipality can create a conducive environment. Furthermore, the strategy emphasises the formalisation of the informal sector. Formalisation is seldom the main problem or ideal outcome, and the focus should rather be on the potential jobs in the informal sector (Fourie, 2018). A further problem is the inhibiting role of government regulations (mostly local government-related) (Turok & Scheba, 2020). These concerns do not receive attention. During one of the interviews, a businessperson said, 'more can be done to make business operations easy'. He referred specifically to the time it takes to change land uses.

The third pillar is innovation and R&D. The strategy sees a market for new products and services. It states that 'the municipality should encourage innovation, research and development by introducing development specialist centres that assist entrepreneurs with designing products and improving operational processes with the latest available technology' (Matjhabeng Local Municipality, 2019, p. 167). With a campus in the area, the Central University of Technology should play an important role in R&D. The strategy proposes establishing a science park and ICT infrastructure to facilitate this idea. There is very little evidence of relatedness or a history of working with technology. Yet there could be some value in pursuing this direction.

The fourth pillar is an enabling business environment. The strategy wants to create 'an environment where the government set standards and legislation, while not inhibiting reasonable business activities' (Matjhabeng Local Municipality, 2019, p. 168). This pillar's three main objectives are improved ease of doing business, training and skill development, and good governance. Each of the objectives has a set of projects that, among other things, include simplifying business regulation. However, the strategy provides little detail and does not refer to the slow processing of land-use applications as one example of the *difficulty* of doing business. The municipality has received only one unqualified audit in the past ten years, but the strategy does not view this as something that requires urgent attention.

The final pillar is tourism. Four projects are envisaged: the development of the Sand River Route, the Film Incubation Hub, the development of a travel guide and revitalising the Phakisa Raceway. Most focus has been on the film incubation hub, but implementation has been slow. Renovation of the raceway (which requires large-scale maintenance) seems unlikely.

7.5 Discussion

Here, we discuss the strategy described earlier in light of the main points we identified in the literature review.

Changing and diversifying an economy is difficult, and even more difficult in secondary cities that are not high on the national policymakers' and bureaucrats' agendas. There is very little indication from our case study that design-based strategies have created new economic paths or will do so in the future. Change has been evolutionary and mainly unrelated to design-based strategies. Still, there is evidence of slow change or the development of an equilibrium or equilibriums. For example, despite being a relatively new town established after the Second World War, Welkom has developed a higher-order regional service function. This is because the 1960 strategy focused on the agricultural sector, which helped develop the Welkom Fresh Produce Market. A thriving rural hinterland has supported the economy of Matjhabeng. Other examples of the regional services function are the small manufacturing industry linked to the agricultural sector through food production, the public and private hospitals with medical specialists, a range of schools and educational organisations with boarding facilities and the decentralised government services. The lack of direct data and the sectoral approach to understanding the economy mean that the consultants and the municipality are not aware of the new equilibrium. The determination to find a magical replacement for mining blinds the planners to the value of the regional services function.

There is ample evidence of path dependency in the Free State Goldfields. Three historical aspects hold back development. A fair share of the mining economy is managed in Johannesburg. This means that the Free State Goldfields does not benefit from mining beyond the immediate mining processes. The literature shows that where cities or regions have managed to diversify their economies, substantial levels of relatedness are present. In the Free State Goldfields, the only project related to gold mining has been the jewellery industry.

The second is the planning system. The towns of Welkom, Virginia and Allanridge developed on garden city principles in the 1950s. Wide roads with a large median strip often with plants and trees, suburban shopping complexes and low densities were common. This required a strict land-use regulation process. To manage decline and shrinkage, higher densities and flexible land-use systems are crucial. Overcoming low densities and dealing with these historical planning realities continues to be a stumbling block in rethinking the city.

The third of the development paths that the Free State Goldfields unthinkingly follows is the false sense of prosperity that mining brings. Despite 30 years of economic decline, the belief in the area's prosperity remained evident in the responses of the people we interviewed. We often heard comments like 'we only need one big investment' or 'the mining will revive if big companies go'. Deacon et al. (2018) found a similar attitude in Canada. Local government officials and business people in the Goldfields, fixated on their new economic development plan and the strategy described earlier, scarcely acknowledge the decline. Nobody wants to acknowledge defeat, but accepting the reality is essential for understanding the area's future.

Plan-making has dominated responses from the Free State Goldfields, apart from some attempts between 1990 and 2004 to create networks, which did help to create a manufacturing sector with global links. However, the most recent plan reinforces the focus on plan-making. For example, the new plan assigns many of the local government's responsibilities to organisations outside the municipality (e.g. the Department of Economic, Small Business Development, Tourism and Environmental Affairs of the Free State Provincial Government). There is an inherent assumption that funds exist to restructure the local economy and that these funds must be allocated by other spheres of government to Matjhabeng. The plans overemphasise activities outside the municipality's control and its local governance structures, such as allocating a special economic zone (SEZ) to the area.

An emphasis on plan-making is often built on assumptions. It assumes that a market exists and that an effortless redirection of economic activity to Matjhabeng is possible. The plan proposes projects without analysing value chains or markets. The idea that Matjhabeng can attract a steelmaking industry does not consider the realities of this industry in South Africa. Currently, the steel industry is operating below capacity, cheap steel imports have affected local steelmaking efforts negatively, one of the largest steelmakers has closed down in Emalahleni in the past decade, and most steelmaking operations are located at the coast to minimise the cost of land transport (Marais et al., 2016, 2021). Relocating or starting a petrol refinery would face many of the same obstacles.

The big ideas envisaged in this plan (SEZs, cargo hubs, steelmaking and petrol refineries), as in many of its predecessors, obscure other possibilities. Although big plans can create local buy-in towards a new economic future, they prevent local planners from looking inward and they ignore the need for relatedness. One of the problems with big plans is that they create dependencies on other institutions for funding and implementation. Furthermore, they underplay local responsibilities. We have already noted that, despite emphasising good governance, the Matjhabeng Local Municipality has produced only one clean audit in the past decade. A clean audit is an essential sign of good governance, but it does not get attention in the plan. More problematic is the poor state of municipal finance, which means that Matjhabeng does not have sufficient municipal funds to allocate to developing an alternative economic future. The current focus ignores the substantial growth in the regional services function over the past three decades. Building this function could mean finding ways to change land use swiftly for services like boarding facilities, guest houses and retail. It would require the municipality to put into practice the improvements it does have control over, rather than being distracted by big ideas.

Another concern about the new economic plan is that an outside consultant developed it, but it must be implemented by local government officials. This may be one of the reasons why it ignores aspects over which the municipality does have control. For example, the plan does not take into account how quickly the municipality can process land-use applications. Respondents said that a simple subdivision takes up to three years to complete. Another example is the plan's acknowledgement that the backlogs in infrastructure hamper long-term growth. Infrastructure and planning is a municipal function, but the plan gives it little attention other than acknowledging it as a stumbling block. The plan does not address the poor state of municipal finance. A good plan and its institutionalisation depend on a commitment from the plan-makers. Overall, the plan makes little reference to the local government functions and efficiencies required to implement the plan.

Finally, there is ample evidence of social disruption because of mine closure. Sesele et al. (2021) and Sesele and Marais (2023) have noted the effect on women. Increased crime is also related to the illegal gold industry and prevents long-term economic development after mining.

7.6 Conclusion

Development planners in the Free State Goldfields have taken various design-based approaches over the past 60 years, but none have succeeded in diversifying the economy, although their contribution to a small manufacturing sector should be acknowledged. These approaches all create the impression of achieving fast growth, but the plans have not been implemented. The Goldfields is held back by a plan-making process that ignores implementation, does not recognise existing economic equilibriums and undervalues its shrinking role. Most seriously for the Goldfields, such planning ignores the area's regional services function and the value of encouraging local responses to this reality.

The question is what the future holds. Most predictions show that the area will lose most of its mining employment over the next ten years (Denoon-Stevens, 2019). Of course, this prediction depends on the gold price. New investments in gas are unlikely to create the same amount of employment but would increase local GVA figures. New industries are planned, but they are unlikely to take off for the reasons outlined earlier. Yet the area has already reached economic equilibrium through its regional services function. A related issue is the importance of acknowledging decline and the fact that the economy is shrinking and using this as a platform to build a new economy. Part of the problem is that the Matjhabeng Local Municipality's planners still count on economic growth.

References

Atienza, M., Lufin, M. & Soto, J., 2021. Mining linkages in the Chilean copper supply network and regional economic development. *Resources Policy*, 70, p. 101154.

Binns, T. & Nel, E., 2001. Gold loses its shine: Decline and response in the South African Goldfields. *Geography*, 86, pp. 255–260.

Binz, C., Truffer, B. & Coenen, L., 2016. Path creation as a process of resource alignment and anchoring: Industry formation for on-site water recycling in Beijing. *Economic Geography*, 92(2), pp. 172–200.

Bolowana, G., 2013. Matjhabeng summit aims for green economy. *The New Age*, 6 December, p. 9.

Boschma, M., 2017. Relatedness as driver of regional diversification: A research agenda. *Regional Studies*, 51(3), pp. 351–364.

Cortinovis, N., Xiao, J., Boschma, R. & van Oort, F., 2017. Quality of government and social capital as drivers of regional diversification in Europe. *Journal of Economic Geography*, 17(6), pp. 1179–1208.

Da Silva, M., 1999. The jewellery industry in South Africa. *Urban Forum*, 10(2), pp. 217–234.

Davenport, J., 2013. *Digging Deep: A History of Mining in South Africa*. Johannesburg: Jonathan Ball Publisher.

Deacon, L., Van Assche, K., Papineaua, J. & Gruezmacher, M., 2018. Speculation, planning, and resilience: Case studies from resource-based communities in Western Canada. *Futures*, 104, pp. 37–46.

Denoon-Stevens, S., 2019. Matjhabeng: Planning in the face of the Free State Goldfields decline. In Marais & Nel, V. (eds.), *Space and Planning in Secondary Cities: Reflections from South Africa*. Bloemfontein: SUN Press, pp. 115–140.

Fourie, F., 2018. *The South African Informal Sector: Creating Jobs, Reducing Poverty*. Pretoria: HSRC Press.

Global Insight, 2020 *Rex Explorer*. www.ihsmarkit.co.za/Products/pdf/IHS% 20Regional% 20eXplorer.pdf

Li, H., Lo, K. & Wang, M., 2015. Economic transformation of mining cities in transition economies: Lessons from Daqing, Northeast China. *International Development Planning Review*, 37(3), pp. 311–328.

Marais, L., 2013a. The impact of mine downscaling in the Free State Goldfields. *Urban Forum*, 24, pp. 503–521.

Marais, L., 2013b. Resources policy and mine closure in South Africa: The case of the Free State Goldfields. *Resources Policy*, 38, pp. 363–372.

Marais, L. & De Lange, A., 2021. Anticipating and planning for mine closure in South Africa. *Futures*, 125, pp. 102669.

Marais, L., Lenka, M., Cloete, J. & Grobler, W., 2016. Emfuleni. In Marais, L., Nel, E. & Donaldson, R. (eds.), *Secondary Cities and Development*. London: Routledge, pp. 83–100.

Marais, L. & Nel, E., 2016. The dangers of growing on gold: Lessons from the history of the Free State Goldfields, South Africa. *Local Economy*, 31(1–2), pp. 282–298.

Marais, L., Nel, V., Rani, K., van Rooyen, D., Sesele, K., van der Watt, P. & du Plessis, L., 2021. Economic transitions in South Africa's secondary cities: Governing mine closures. *Politics and Governance*, 9(2), pp. 381–392.

Marais, L., van Rooyen, D., Nel, E. & Lenka, M., 2017. Responses to mine downscaling: Evidence from secondary cities in the South African Goldfields. *The Extractive Industries and Society*, 4, pp. 163–171.

Martinez-Fernandez, C., Wu, C.-T., Schatz, L., Taira, N. & Vargas-Hernández, J., 2012. The shrinking mining city: Urban dynamics and contested territory. *International Journal of Urban and Regional Research*, 36(2), pp. 245–260.

Matjhabeng Local Municipality, 2019. *Local Economic Development Strategy*. Welkom: Matjhabeng Local Municipality.

Nel, E. & Binns, T., 2002. Decline and response in South Africa's Free State Goldfields: Local economic development in Matjhabeng. *International Development Planning Review*, 24(3), pp. 249–269.

North, D.C., 1990. *Institutions, Institutional Change and Economic Development*. New York: Cambridge University Press.

Page, D., 1969. The O.F.S. goldfields: Planning for an emerging metropolis. Unpublished master's dissertation, University of Stellenbosch.

SACN (South African Cities Network), 2020. *Profiling Intermediate Cities in South Africa*. Johannesburg: SACN.

Sesele, K. & Marais, L., 2023. Mine closure, women and crime in Matjhabeng, South Africa. *Geographical Research*, 61, pp. 18–31.

Sesele, K., Marais, L., van Rooyen, D. & Cloete, J., 2021. Mine decline and women: Reflections from the Free State Goldfields. *The Extractive Industries and Society*, 8(1), pp. 211–219.

Turok, I. & Scheba, A., 2020. Strengthening township economies in South Africa: The case for better regulation and policy innovation. *Urban Forum*, 31, pp. 77–94.

Welkom Municipality, 1968. *Welkom: Capital of the Orange Free State Goldfields*. Johannesburg: Felstar Publishers.

Welkom Municipality, 1989. *Welkom Structure Plan*. Welkom Municipality.

8 Kleinzee

Looking for a new dawn amidst the diamond dust

Anmar Pretorius and Derick Blaauw

8.1 Introduction

The history of Kleinzee is intrinsically linked to the mining of alluvial diamonds, which began in the 1920s under De Beers but went into decline in the 2000s when the diamonds began to run out. This small coastal town is situated in the Nama Khoi Local Municipality in the Northern Cape's remote Namaqualand, famed for its spring flowers and spectacular desert landscapes (see Figure 8.1).

Kleinzee originated as a prototypical company town to supply the labour needs of De Beers Consolidated Mines. This type of town was common in South Africa up to the 1990s (see Marais et al., 2018, for a discussion of the situation of company towns globally). Kleinzee, for De Beers employees only, was fenced and had security measures such as an official checkpoint at the town's entrance gate. De Beers was solely responsible for the town's development and infrastructure and owned 320 km of coastal land that includes Kleinzee and stretches as far as Port Nolloth.

The remaining residents reflect fondly on the history of their town. A 2011 article from the *Mail & Guardian* (Stilwell, 2011) gives a wealth of detail. It notes that at the peak of mining in the 1980s, more than 3,000 people were employed in Kleinzee, and its population was about double that number. Medical facilities for the residents included a fully equipped hospital with doctors, dentists, radiographers, physiotherapists and a maternity unit. Visiting specialists performed surgery, including cosmetic plastic surgery, and a plane was on standby for medical emergencies. Recreational amenities, funded by the company, included an 18-hole golf course, two Olympic-size swimming pools, and various sporting and social clubs for activities such as camping, horseback riding, and arts and crafts. In 1981, there were three schools with 651 students. Carstens (2001) notes that the primary school was one of the first in South Africa to be bilingual (Afrikaans and English). It was desegregated by admitting coloured and black children in 1988, when De Beers announced that all forms of apartheid would be abolished in the town. This process was completed in 1991. This meant that the suburbs and schools were not planned on the basis of race any more.

Life for the residents was indeed good and well worth reminiscing about today, as one of our interviewees poetically explained:

DOI: 10.324/9781003403326-8

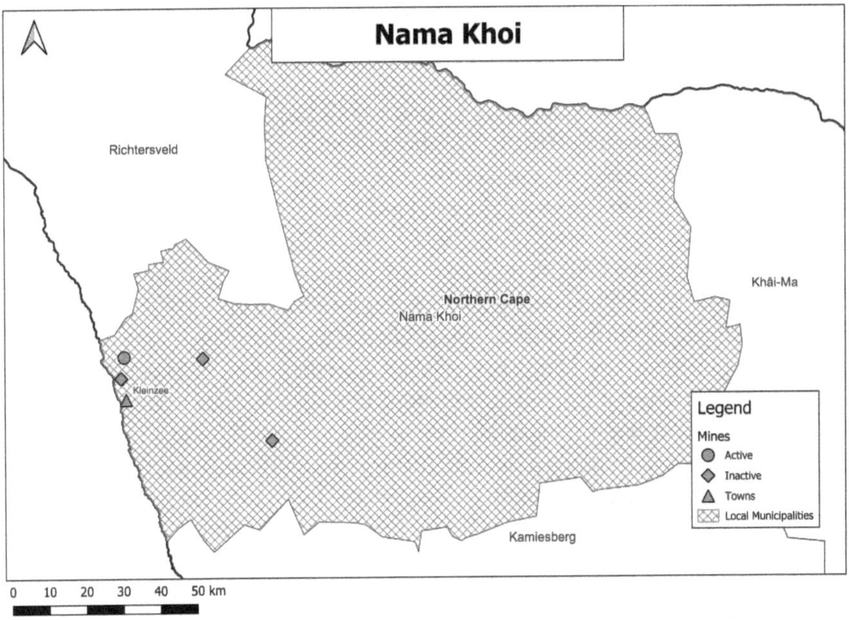

Figure 8.1 Mines and settlements in Kleinzee

> The residents lived well in Kleinzee. It was in effect like moonshine and roses, if you will. It was one of the most picturesque prosperous towns you can imagine. The roads were of the best, as were the houses. People were proud to live and work there for De Beers.

The boom inevitably came to an end. Mine decline brought many changes. In 2008, De Beers applied for Kleinzee to be declared a public town and for it to be incorporated into the Nama Khoi Local Municipality. This process of 'normalisation' (declaring a company town a public town) is not uncommon in South Africa when a mine goes into decline (see Van der Watt & Marais, 2019 for a discussion of the issues involved). Consequently, Kleinzee was proclaimed a public town in 2012, the year that De Beers stopped official operations. In May of 2011, the mine had been sold to Trans Hex. This did not entail taking over responsibility for the town. In 2020, De Beers was still delivering some municipal services to the residents, and large portions of land were still owned by De Beers. Local and regional communities are struggling to deal with the economic and social consequences of declining mine activity.

The socio-economic impact of the official closing of the De Beers diamond mine is felt by the remaining residents. The population fell steeply from around 4,000 in the early 1990s to 728 in 2011 (Stats SA, 2014). The town has fallen into neglect since 2013, when De Beers auctioned off most of its properties – sports fields, clubhouses and hundreds of houses. Problems quickly began to mount as the municipality simply did not have the funds to maintain the existing infrastructure. A respondent we interviewed said:

> They [the municipality] struggle to make ends meet and deliver basic services to the community. That is all. There is nothing left to do anything else with. As a result, the roads are bad, the recreational club has closed down and the town has become dilapidated and run-down.

As is the case with many South African mines that have closed or are in decline, the environmental damage has only partly been remedied, with significant rehabilitation remaining to be done. The damage remains a long-term legacy and dependency that will haunt the town. Shops have closed, and the hospital where complicated operations were once performed now renders only basic services to the community. Today, only one of the former three primary schools is still operational. Kleinzee Primary School is currently a public school that specialises in special needs education. One Olympic-size swimming pool is empty, and a restaurant reportedly uses the other to keep its fish stock alive before serving it at mealtime. The dilapidation of this infrastructure is the result of the broken interdependency between the mine and the community. When mining was prospering in Kleinzee, De Beers depended on the miners for their labour, and the miners depended on De Beers for the maintenance of the town's infrastructure and many amenities. When the diamond supply diminished, De Beers no longer needed the miners or the town, whereas the miners and the town still needed De Beers. The miners and the town are the losers.

The chapter describes the social disruption in the form of rapid changes and breakdowns in the social life of the Kleinzee community because of the decline in mining activities. We discuss the town's history and follow its decline from a once prosperous and safe home to its current poor state. The departure of De Beers was a severe economic and social shock, and crime, an extreme form of social disruption, has followed as an aftershock. Crime remains a primary hindrance to the many positive efforts to develop an alternative economic base, such as tourism, and the chapter highlights this through some voices representative of its community.

The views and experiences of the residents were obtained through in-depth, semi-structured interviews with five people who were stakeholders and role-players in the community. We used snowball sampling to find people willing to be interviewed. The potential interviewees were contacted via telephone, and the purpose of the research was explained to them. They then had the opportunity to accept or decline the invitation.

An in-depth interview was held with the manager of the local community radio station and a former small business owner. We interviewed the owner of a business currently operating and spoke to someone who grew up in the area as well as a former council member. Our attempts to interview people from the municipality, the school and De Beers were not successful. Our five interviews did, however, provide in-depth information and insight that were indispensable in forming the picture we present in this chapter. We combined this approach with background research in related literature and theory, local government documentation and statistical information from Statistics South Africa and other quantitative sources.

Table 8.1 Namakwa District Municipality: sectoral contribution to gross GDP (in %)

	Agriculture	*Mining*	*Electricity*	*Construction*	*Trade*	*Transport*	*Finance*	*Community services*
1996	6.44	48.62	1.62	1.77	8.15	5.14	7.18	10.93
2000	5.96	57.37	1.05	1.09	7.09	4.08	6.35	8.70
2004	6.42	53.47	0.94	1.23	8.29	5.02	6.65	9.45
2008	8.39	44.18	1.02	1.83	9.96	6.24	7.98	11.33
2012	8.31	39.56	0.86	2.11	10.82	6.83	8.77	12.99
2016	8.63	39.15	0.28	2.21	10.79	7.10	9.15	13.04
2019	8.45	38.82	0.20	2.03	10.44	6.99	9.20	13.17

Source: Authors' calculations using data from IHS Markit Regional eXplorer (2021) www.ihsmarkit.co.za/

Note: Percentages are not reported for all the economic sectors or for all years, although the data are available annually. As not all sectors are reported, the contributions do not add up to 100%.

8.2 Declining economic role of mining

Globally, diamond mining activity has declined since the mid-1990s. The Parliamentary Monitoring Group (2010) reported in 2010 that the Kleinzee mine had about 1.7 million carats left for mining activities (i.e. 100,000 carats for 17 years). De Beers decided not to continue their own mining operations but started the process of selling the Kleinzee diamond mines to Trans Hex (operating as West Coast Resources) in 2011. Trans Hex would continue with mining and rehabilitation activities. However, in 2019, Trans Hex decided to liquidate West Coast Resources, citing the unprofitability of diamond mining due to lower diamond prices and the increase in supply of laboratory-grown diamonds (*Business Day*, 2019). Table 8.1 shows the Namakwa District Municipality's economic base between 1996 and 2019, with the sectoral contributions to gross GDP.

Besides the De Beers diamond mines to the north of Kleinzee, the Namakwa District Municipality also hosts copper mines near Springbok and Okiep. Nevertheless, the declining role of mining is evident from the percentages in this table. In 1996, mining contributed 48.6% to gross GDP. This reached 57.4% in 2000 and then steadily declined to 38.8% in 2019. The contribution of all the other sectors, except electricity, has increased, though with a small decline for some of them in 2019. The role of tourism, not shown in the table, is difficult to isolate, as it is partly reflected in trade, transport and community services. The growth in these three sectors, however, may indicate a growing tourism industry in the municipality.

8.3 Demographic change

Table 8.2 shows a general increase in the population of the Nama Khoi Local Municipality since 1996.[1] The ratio of male to female population increased slightly between 1996 and 2011, but not to the extent usually observed in mining towns, where inward migration increases the share of the (younger) male population. Notably, unemployment increased markedly between 1996 and 2016. Yet living

Table 8.2 Population trends in Nama Khoi Local Municipality

	1996	*2001*	*2011*	*2016*	*2018*
Population	43,841	44,900	47,041	46,512	
Formal dwelling	8,779	9,695	12,489	13,600	
%	20.03	21.59	26.55	29.24	
Unemployment	15%	28%	26%	28%	26%
Male	21,446	22,099	23,215	22,835	
%	48.92	49.22	49.35	49.10	
Female	22,395	22,801	23,826	23,677	
%	51.08	50.78	50.65	50.90	

Source: Nama Khoi Municipality (2019)

Table 8.3 Population trends in Kleinzee, 2001–2011

	Kleinzee 2001	*Kleinzee 2011*
Population	2,712	728
Household income below R19,600 pa	14%	14%
Individual monthly income below R1,600 pm	29.04%	28.51%
Population employed	86.03%	50.41%
% of the population working in the formal sector	79.07%	42.56%
% of the population living in informal housing	0.19%	0%
% of the population living in fully paid own dwelling	0.39%	0%
Water, electricity, flushing toilets in the house	100%	100%
Population 20+ not matric	56.99%	43.53%
Population with tertiary education	18.82%	5.29%

Source: Stats SA (2014)

conditions have improved, with more residents living in formal dwellings in 2016 than in 1996. In another positive development (not shown in the table), the share of the population older than 20 without schooling decreased from 11.1% to 1.5%, and the share with a higher education qualification increased from 4.7% to 7.2%. Official statistics (also not shown in the table) present a picture of an ageing population over the 20 years from 1996 to 2016, with a decrease in the age group 0–19 and an increase in the population older than 65 (Nama Khoi Local Municipality, 2019).

Population trends in the town of Kleinzee present a different picture. Table 8.3 shows a population decline of almost 73% between 2001 and 2011.[2] Official employment dropped dramatically, reflecting the severe impact that mine decline has had on the town. Yet average household and individual monthly income, in nominal terms, did not decline to the same extent. The share of the population without matriculation increased, and the share with tertiary education decreased. We can reasonably conclude that a large part of the population loss in Kleinzee can be attributed to highly skilled individuals who left the town when mining activities declined and were eventually halted. This loss of skilled people is an example of social disruption caused by mine decline. It stands in contrast to the social disruption caused by the rapid inflow of people when a mine opens, which has been the main focus of the literature to date.

Table 8.4 Number of reported crimes per category

Year	*Malicious damage to property*	*Theft out of/from vehicle*	*General theft*	*Property-related crimes*	*Drug-related crimes*
2006	3	1	4	10	1
2007	4	2	6	7	3
2008	1	1	1	2	2
2009	0	1	2	5	1
2010	3	1	8	5	1
2011	1	0	1	5	4
2012	1	0	8	2	5
2013	7	2	12	7	7
2014	1	1	6	4	2
2015	0	0	9	7	7
2016	4	0	10	15	4
2017	4	3	14	18	12
2018	4	2	13	10	10
2019	14	5	20	28	28
2020	20	2	36	28	4
2021	11	3	20	10	1

Source: Institute for Security Studies Crime Hub (2022)

Note: The years run from 1 April to 31 March. The database categorises 'Total property related crimes' as the sum of the following: burglary at non-residential premises, burglary at residential premises, theft of motor vehicles and motorcycles, theft out of or from motor vehicles, stock-theft. We listed 'theft out of or from vehicle' on its own as well because of the increase in reported cases. Even though there were not many, the residents had not experienced this kind of crime before, and it increased their anxiety and feelings of not being safe in Kleinzee anymore.

8.4 Crime

In contrast to some findings in the literature for the Global North (O'Connor & Ruddell, 2021), Marais et al. (2022) found that mine decline and closure lead to social disruption in the form of an increase in crime. Table 8.4 shows crime statistics for Kleinzee from 2006 to 2021, from the Institute for Security Studies Crime Hub (2022),[3] as reported per police station by the South African Police Service. The table does not show some serious offences, such as murder, sexual offences and assault, as these were very rare and did not show any particular trend.

There has been a marked increase in cases reported, particularly since 2016. The decline towards the end of the period may be due to the various COVID-19 lockdown levels. However, given the declining population in Kleinzee, the increase in reported cases points towards an even larger increase in cases per capita. The drug-related and other crimes listed in this table may indicate increased activity by non-residents. The trends in the table were echoed by an interviewee who stated that residents no longer feel safe in Kleinzee. This interviewee also observed that the official security gate at the entrance to the town was ineffective in preventing the influx of illegal miners and other elements vandalising properties.

8.5 Zama zamas and disruption[4]

Kleinzee shares a notable characteristic with many other communities where mining has declined in scope and intensity: the emergence of artisanal and illegal mining. In South Africa, informal and illegal miners are referred to as *zama zamas*, meaning loosely 'ones who try their luck' in isiZulu. The presence of zama zamas contributes to higher crime levels (and other social disruption), particularly as many of them operate in syndicates (Marais et al., 2022).

The diamond mining sector's steady decline in South Africa and the scaling down of diamond mining in Kleinzee have left a legacy of chronic unemployment in Namaqualand. The pre-existing interdependency between the community and De Beers has shaped the situation since the decline of mining. However, the existing skills among the former miners (gained as a result of an erstwhile mutually beneficial interdependency with De Beers) now create opportunities for a growing number of them, as well as for other desperate young men and women from neighbouring countries who are willing to endure hardships and hazards as they illegally eke out a living in abandoned mines in the area. See Trenchard (2022) for a vividly illustrated account of the *zama zama's* lives.

The hardships and dangers for the estimated 1,000 active illegal miners in the Namakwa District Municipality are real (Netwerk24, 2018). The media has reported several fatalities in the past decade in some of the districts with deserted mines. For example, in May 2012, ten people died in an accident, and similarly, in October 2021, five illegal miners were presumed to be dead in the aftermath of a collapse of unstable tunnels in the unused Nuttabooi mine in the Nama Khoi Local Municipality.[5]

The policy response to the rise in the number of illegal miners in Kleinzee and the broader Namakwaland shows definite signs of what North (2005) would term path dependency, implying that the historical dependence on the mining sector has shaped the nature of the policy response to the illegal mining phenomenon as well as the speed thereof. For example, in 2018, the government announced a breakthrough in negotiations with illegal miners (Netwerk24, 2018). In meetings with the deputy minister of mineral resources, Godfrey Oliphant, it was agreed at the end of November that activities would be suspended until January 2019, when the government would issue permits to formalise and legalise the activities of the illegal miners. The deputy minister undertook to return in two weeks to give feedback to the miners regarding the progress made with the issuing of permits (Netwerk24, 2018). Four years later, the permits had not yet materialised, and the legal status of the *zama zamas'* mining activities remains uncertain. The illegal miners blame De Beers for the situation in which they find themselves. Journalist Trenchard (2022) recorded the words of one of them:

> De Beers has taken all the diamonds out of Namaqualand and now we're back in poverty. . . . The companies left us like they found us – with nothing. But if we just sit there thinking about it, our kids will die. So we, the diggers, come in here and take the scraps.

This is further evidence of how interdependencies can lead to distrust and conflict, as Marais et al. (2021) have argued. Unfulfilled promises like the failure to issue permits have exacerbated the distrust and anger, rendering the already fragile relationships even more precarious.

The authorities, in turn, are trying to use traditional law enforcement methods to deal with the problem. For example, in March 2021, a man was arrested in Port Nolloth for possessing 254 illegal diamonds (News24, 2021). Between 30 October 2019 and 4 November 2019, 15 people were arrested for illegal mining activities near Kleinzee. Mining tools and gravel containing diamonds were confiscated in the process (Netwerk24, 2019a). In January 2019, 1,800 artisanal miners were evicted following an interdict in the Northern Cape High Court. The Nama Khoi municipality is currently engaged in litigation, which could result in damages and costs in the region of R300,000 being awarded against the Council if the claimants' case is successful. In addition, a court review is currently in process over the eviction of illegal miners from Kleinzee by De Beers. The financial burden could be catastrophic for a local municipality that is struggling to make ends meet, according to one of the community members interviewed. There are clear issues of goal dependency and steering (Beunen & Van Assche, 2021, and see Chapter 2) at play in Kleinzee. Local governments, facing legal and financial difficulties, are struggling to adapt to decline and closure in the diamond mining sector and to respond to the consequent illegal mining in a timely and flexible manner, while the national government is dragging its feet.

The delay, one of our interviewees said, has made the miners even more abrasive in their conduct. The remaining residents of Kleinzee are voicing their concern. Trenchard (2022) reports residents saying the influx of illegal miners from other areas of South Africa and from neighbouring countries has increased crime in their once-safe town. He quotes a Kleinzee restaurateur, saying:

> *Zama zamas* aren't welcome here. . . . They have knives, they have guns, they're gangsters. They can come and buy things here, but afterward they must leave. We don't want them here.

From our interviews with community members, it became apparent that the ongoing issue of illegal mining and the increase in crime make it even more difficult to realise the admirable goal of expanding tourism as an alternative economic base for Kleinzee. Attempts have been made to market the town as a tourism destination. But our interviewees expressed sentiments like the following:

> The things that make a town a tourism attraction are now being run down, such as the infrastructure and roads. The crime is getting worse, partly due to the influx of the *zama zamas*. These illegal miners have taken away the hope of the town. For too long, nothing has been done about this.

They complained about the absence of plans to deal with the situation:

> The government has waited way too long to do anything about this issue and as a result the problem has now escalated beyond the ability of the government to control it. Issues such as increased crime, vandalism and the destruction of infrastructure have become a major problem.

Whereas Kleinzee was once the safest place imaginable, with security and a checkpoint at the gate, the opposite is now becoming true. The fallout after De Beers sold their mines to Trans Hex and the subsequent mining decline, coupled with the deteriorating crime situation, threaten the endeavour to develop a viable tourism industry as a future economic base. More crime implies fewer visitors, worsening economic conditions, and increasing crime levels again – a vicious circle that will be difficult to break.

A related problem for the remaining residents is the drop in property values. Many people who bought a property here are now stuck with it, watching its value decrease further as crime and vandalism increase. This leaves the town's future and long-term development, transitioning away from dependence on mining, in obvious jeopardy. Without a holistic and speedy approach to dealing with this problem, the prospect of a new dawn arising from the dust is fading.

8.6 Quo Vadis? Hopes for a new dawn

The past decade has seen several initiatives and proposals for Kleinzee to break its dependence on mining. The local government documentation that we examined, such as the municipality's integrated development plan (IDP), proposes such things as upgrading the hospital, constructing wind farms, expanding the local airport and developing a world-class sports training facility. One ambitious proposal is to turn a redundant hostel and workshops into a prison, consisting of a rehabilitation centre and training facilities. De Beers made a pre-feasibility contribution to support local communities. National and provincial government support was sought, with an envisaged time frame of 2011 to December 2012. But nothing has come of this.

De Beers did initiate certain social responsibility projects as part of its exit policy. One of the more successful projects is Diamond Coast Aquaculture, involving the collection of abalone and seaweed (De Beers, 2009). The abalone farm currently produces approximately 120 tonnes of abalone per year, which, with the support of the Viking Aquaculture distribution network, is marketed under the Sea Harvest brand (Netwerk24, 2019b; Viking Aquaculture, n.d.). Another is the De Beers Zimele Kleinzee Hub, which trains and supports owners of small and medium-sized enterprises (Free to Grow, 2018).

We heard from community members that several people had been helped by the De Beers entrepreneurship assistance programmes. Residents were given loans to buy buildings and premises, but with the town's deterioration and the loss of its economic base, the buying power disappeared, and most of these businesses have been closed or scaled down to small home industries, for example baking cakes for individuals rather than operating a bakery. An interviewee said:

> De Beers gave loans to these people while fully knowing that they would close the mine down. They must have known what the impact would be on these fledging businesses. In some cases, there are disputes between the former entrepreneurs and De Beers regarding paying back these loans, for example.

This opinion contrasts with the positive picture of De Beers assisting local people in entrepreneurship ventures (Free to Grow, 2018), as portrayed in the popular press.

Residents do not support all the proposals for increased economic activity. The possibility of seismic exploration for oil and gas has been met with fierce resistance from local fishermen and other groups. In most cases, the argument is that insufficient meaningful consultation has taken place between the companies intending the exploration and the people who would be most affected by the activities (Netwerk24, 2022). The jury is still out on the eventual outcome of the intended exploration and the supposed benefits for the Kleinzee community.

If Kleinzee is going to see a new dawn, it will probably come from the town's own efforts, without any reliance on mining. Unlike some towns that are the subject of other case studies in this book, Kleinzee has several advantages to build on and expand on. Its location is one such advantage. The inherent appeal of its rural setting next to the sea naturally predisposes it to developing its tourism sector (World Tourism Organization, 2018) and its eco-tourism industry in particular. There are several shipwrecks along the coast that can serve as tourist attractions (Van Dyk, 2021). The natural environment offers the possibility of 4x4 routes and fishing trips. The municipality recognised this in its 2020 IDP, where it envisages marketing the 'small town coastal character' and promoting the town as a holiday destination with coastal front holiday properties and investment opportunities, particularly catering for the wilderness tourism market (Nama Khoi Municipality, 2019).

Although residents agree that this vision has significant merit, they see the increasing crime as a stumbling block. They fear that rising criminal activity will cause visitors to avoid the town and book their stay further along the coast. The town might not benefit from the overnight stays that are the backbone of recent tourism development. It is therefore vital to deal with the crime if tourism development, as outlined in the IDP, is to become a reality.

We heard of entrepreneurs who have kept going in spite of the obstacles. There are, for example, currently 11 guest houses in Kleinzee. The sense of community is enhanced by activities and endeavours such as the local community radio station, which attempts to give hope to the community and offers an opportunity for the remaining residents to reminisce about the days when things were better. It features stories of hope to keep the community going despite the hardships they are experiencing.

8.7 Conclusion

This chapter explored the contrast between the town's prosperous past and its socio-economic decline today. We looked at the hardships experienced by the remaining residents of Kleinzee and their efforts to take the town forward and promote it as, for example, a rural tourist destination.

Kleinzee originated as a settlement to support De Beers' diamond mining operations on the coast of Namakwaland. The town prospered during the boom. But the diamond mining bust has brought the troubled times and hardships shared by most erstwhile mining communities in similar circumstances. Schools have closed, the population has dwindled, and the economic base and infrastructure cannot be maintained, although the municipality does its best to provide the basic services with a shrinking tax base and income. Several factors hamper attempts to refocus economic activities on industries such as tourism. Foremost is social disruption manifesting as rising crime, closely related to the influx of illegal miners who continue to dig for diamonds in the dust of the mine formerly owned by De Beers. Crime has brought down property values and inhibited progress towards a tourism-based local economy. The national government's current policy inertia in dealing with illegal mining threatens the community's valiant efforts to overcome the mining-induced dependencies. The municipality's inability to maintain the town's infrastructure leads to other forms of disruption and discourages private investment.

The reality facing this community is that it would not have been there had it not been for the discovery of diamonds in the first place. It may have to face the fact that its peripheral location means that many of its idealistic plans for economic recovery and development may not be viable. The community will never be able to match the levels of investment made and maintained by De Beers over many years.

However, Kleinzee's coastal wilderness location gives it an advantage over many other mining towns. Its many natural endowments offer plenty of potential to develop its tourism industry. The remaining residents' resilience, community spirit and entrepreneurship, coupled with their considerable human capital, may enable the town to emerge from the diamond dust by going back to where the town would have been if mining had never started – back to the basics, so to speak.

A fresh start for the broader Kleinzee community could be based on tourism and the recent developments in aquaculture. The existing infrastructure that is in good condition, such as the roads and the airport, and the steady supply of electricity (Kleinzee, unlike the rest of South Africa, has never suffered rolling electricity blackouts), could help to accelerate these and other initiatives. The economic development will probably not be based primarily on the current version of Kleinzee as a town, given the increasing crime and the often illegal occupation of uninhabited houses, but will be more likely to happen along the coastal areas near the town. The road ahead will not be an easy one, but the alternative is one of escalating levels of deprivation and desperation for a town that was once a sparkling gem in the landscape of Namakwaland.

Notes

1 The statistics in Table 8.2 were sourced from the IDP (Nama Khoi Local Municipality, 2019), which includes more recent data. However, the dates do not correspond with the official census years. The IDP figures are probably estimates and that is why the official figures are mentioned in the text and not included in the table. The population figures in Table 8.2, sourced from the IDP, include more recent values, but not for Kleinzee specifically.

2 The data in Table 8.3 are based on official census surveys. These surveys record more variables than the ones reported in the IDP (Nama Khoi Local Municipality, 2019). The most

recent date for which we have census data on Kleinzee specifically is 2011, the last official census date. This is a reasonable time period after the departure of De Beers. So even though the 2011 figures are not recent, they do reflect the impact of the mining decline.

3 Regional Explorer publishes crime statistics per municipality, whereas the Institute provides them per police station. The Institute data are only available from 2006, but we include it here to focus on the town of Kleinzee.

4 Much of our information in this section is drawn from the Afrikaans news source Netwerk24. As Kleinzee is primarily an Afrikaans-speaking area, this is the primary language of the news reports.

5 This was reported by Netwerk24 in an audio clip (in Afrikaans) on 15 October 2021: *Onwettige delwers toegeval in onstabiele diamantmyn* [Illegal miners buried alive in unstable diamond mine]. The link also has a picture that vividly conveys the conditions in which the *zama zamas* work.

www.netwerk24.com/netwerk24/nuus/aktueel/onwettige-delwers-toegeval-in-onstabiele-diamantmyn-20211015

References

Beunen, R. & Van Assche, K., 2021. Steering in governance: Evolutionary perspectives. *Politics and Governance*, 9(2), pp. 365–368.

Business Day, 2019. *Another Blow to Christo Wiese as Trans Hex Liquidates Diamond Mines*. 21 October. www.businesslive.co.za/bd/companies/mining/2019-10-21-another-blow-to-christo-wiese-as-trans-hex-liquidates-diamond-mine/

Carstens, P., 2001. *In the Company of Diamonds: De Beers, Kleinzee, and the Control of a Town*. Athens: Ohio University Press.

De Beers, 2009. *Report to Society 2009: Living Up to Diamonds*. www.debeersgroup.com/~/media/Files/D/De-Beers-Group-V2/documents/reports/library/rts09-full-report-may-2010pdfdownloadasset.PDF

Free to Grow, 2018. *Making Dreams a Reality: De Beers Zimele Celebrates Empowered Entrepreneurs*. https://freetogrow.com/making-dreams-a-reality-de-beers-zimele-celebrates-empowered-entrepreneurs/

Institute for Security Studies Crime Hub, 2022. *Local Crime Statistics*. https://issafrica.org/crimehub/about-us/what-is-the-crime-hub

Marais, L., Campbell, M., Denoon-Stevens, S. & Van Rooyen, D., 2021. *Mining and Community in the South African Platinum Belt: A Decade after Marikana*. New York: Nova Science.

Marais, L., Haslam McKenzie, F., Deacon, L., Nel, E., Van Rooyen, D. & Cloete, J., 2018. The changing nature of mining towns: Reflections from Australia, Canada and South Africa. *Land Use Policy*, 76, pp. 779–788.

Marais, L., Ndaguba, E., Mmbadi, E., Cloete, J. & Lenka, M., 2022. Mine closure, social disruption, and crime in South Africa. *The Geographical Journal*, 188(3), 383–400.

Nama Khoi Local Municipality, 2019. *Integrated Development Plan (IDP) – Second Revision 2019/2020*. www.namakwa-dm.gov.za/wp-content/uploads/2019/04/NDM-Draft-IDP-2019-2020-29-Mrt-2019-final.pdf

Netwerk24, 2018. *Mynbedrywighede tydelik in Namakwa gestaak [Mine Activities Stopped Temporarily]*. www.netwerk24.com/Netwerk24/bedrywighede-eers-gestaak-20181128-4

Netwerk24, 2019a. *Crime Combatted*. www.netwerk24.com/netwerk24/za/kalahari-bulletin/nuus/crime-combated-20191112-4

Netwerk24, 2019b. *Sea Harvest spog met groei [Sea Harvest Boasts Growth]*. www.netwerk24.com/netwerk24/za/weslander/sake/sea-harvest-spog-met-groei-20190313-2

Netwerk24, 2022. *Coastal Blast Row Re Ignites*. www.netwerk24.com/netwerk24/za/weslander/nuus/coastal-blast-row-re-ignites-20220413-2

News24, 2021. *Northern Cape Man Arrested for Alleged Diamond Dealing*. www.news24.com/news24/southafrica/news/northern-cape-man-arrested-for-alleged-diamond-dealing-20210321

North, D., 2005. *Understanding the Process of Economic Change*. Princeton: Princeton University Press.

O'Connor, C. & Ruddell, R., 2021. After the downturn: Perceptions of crime and policing. *The Canadian Geographer*. https://doi.org/10.1111/cag.12671

Parliamentary Monitoring Group, 2010. *Report of the Select Committee on Economic Development on its oversight visit to Port Nolloth, Northern Cape*, 01 June. https://static.pmg.org.za/docs/2010/comreports/100819sceconomicreport.htm

Stats SA (Statistics South Africa), 2014. Mid-year population estimates. *Statistical Release P0302*. Pretoria: Statistics South Africa.

Stillwell, M., 2011. Diamond mines are not forever. *Mail & Guardian*, 18 November. https://mg.co.za/article/2011-11-18-diamond-mines-are-not-forever/

Trenchard, T., 2022. *Diamond Diggers in South Africa's Deserted Mines Break the Law – And Risk Their Lives*. www.npr.org/sections/goatsandsoda/2022/11/27/1132369294/diamond-diggers-in-south-africas-deserted-mines-break-the-law-and-risk-their-liv

Van der Watt, P. & Marais, L., 2019. Normalising mining company towns in Emalahleni, South Africa. *The Extractive Industries and Society*, 6(4), pp. 1205–1214.

Van Dyk, J., 2021. Diamonds are forever: (Re)building the archive of a prototypical mining town. Master's thesis in the Faculty of Arts and Social Sciences, Stellenbosch University.

Viking Aquaculture, n.d. *Diamond Coast Aquaculture*. www.vikingaquaculture.co.za/abalone/farming/

World Tourism Organization, 2018. Tourism for development – Volume I: Key areas for action. *UNWTO*, Madrid. https://doi.org/10.18111/9789284419722

9 Koffiefontein mine downscaling

Socioeconomic and infrastructural consequences

Sethulego Matebesi, Thomas Stewart, Maléne Campbell and Oupa Kale

9.1 Introduction

This chapter describes the socioeconomic and infrastructural consequences of the downscaling of the Koffiefontein mine. It looks at how the livelihoods of residents, retrenched miners and business owners have been affected and compares the current situation with the predictions of a social impact assessment (SIA) done in 2005. The international literature notes that communities are often excluded from mine closure decision-making (see e.g. IISD, 2021; Monosky & Keeling, 2021). This exclusion causes tension between mining companies and communities and leads to social disruption. To investigate the situation in Koffiefontein, we conducted 20 semi-structured interviews with retrenched miners, business owners, municipal officials and community leaders and examined municipal and mining company documents (see Figure 9.1).

The Constitution of the Republic of South Africa (RSA, 1996) requires local governments to provide basic services. Municipalities must develop integrated development plans (IDPs) to show their strategic intent and service provision plan. In mining communities, these IDPs must dovetail with mining companies' social and labour plans (SLPs). The 2020/2021 IDP for Letsemeng Local Municipality, which includes the small mining town of Koffiefontein, emphasises the integral role of communities in participating in socioeconomic development, requiring them to take ownership and protect local infrastructure. Being a mining town is not necessarily an advantage since mining can lead to an influx of people and pressure on the infrastructure. The dovetailing of IDPs and SLPs is thus crucial. Mine downscaling and closure require a similar planning response, but less attention is paid to this. For in-depth analysis and discussion of pressures on mining towns, see for example Bainton & Holcombe (2018), Campbell (2021), Marais, Burger et al. (2022) and Marais, Ndaguba et al. (2022).

9.2 The Koffiefontein mine

The history of the Koffiefontein mine stretches back more than 150 years. It was one of the world's top-producing kimberlite mines and is well known for its high-value diamonds. Nevertheless, over the years, it has opened and closed many times

DOI: 10.324/9781003403326-9

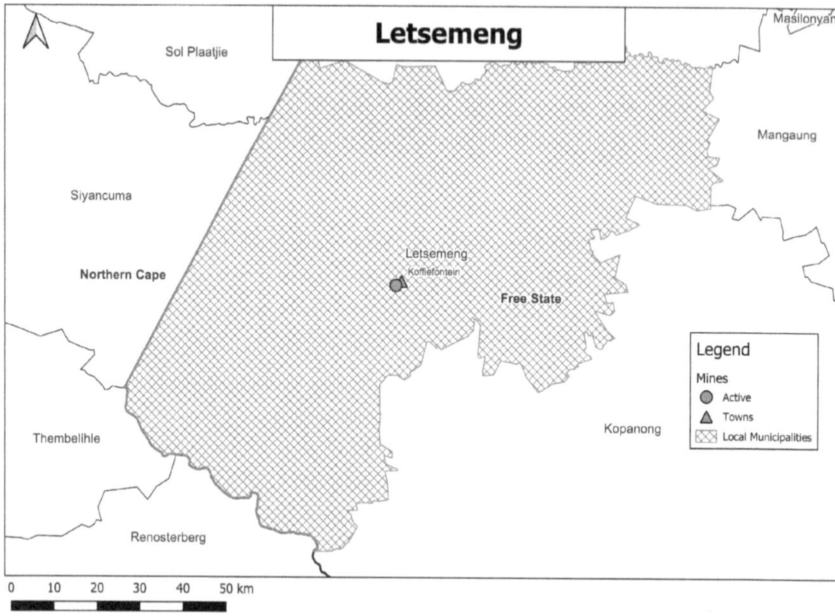

Figure 9.1 Mines and settlements in Letsemeng

(McGill, 1991). Ownership changes and closures were common in the early days (1870–1930). De Beers relinquished its mining rights in 1892 but regained control in 1911 (*Diamond Fields Advertiser*, 1987). The Great Depression forced the mine to close in 1932. It resumed operations in 1950, closed again in 1953 and reopened in 1971. It closed once again in 1982 and then reopened in 1987 (CDS, 2005). In July 2007, De Beers sold the mine to Petra Diamonds (a typical example of a large mining firm selling a declining mine to a small company). In early 2023, Petra Diamonds said it wanted to sell the mine. At the time of writing, it had started reducing its staff and was proceeding with a decommissioning, rehabilitation and closure programme for closure in 2025, should it not find a new buyer in the process.

9.3 Koffiefontein mine social impact assessment, 2005

Mining operations have infrastructural, environmental and social impacts. To improve understanding of the vulnerability and risks associated with these impacts, the Koffiefontein mine submitted an environmental management programme that included an SIA in 2007. The evidence from the SIA showed that the community depended heavily on the mine, not only for employment but also for infrastructure maintenance. The mine had been performing some municipal functions, and the SIA highlighted the risks of mine closure for infrastructure in Koffiefontein.

In 1996, Koffiefontein had a population of 9,758, which increased slightly to 10,402 in 2011 (Stats SA, 2011). According to the SIA, a decline in population size and a change in labour migration patterns would have a moderately negative impact (CDS, 2005). Communities are particularly at risk when local booms and busts lead to rapid demographic changes (Atkinson et al., 2016). Other impacts of closure that were mentioned in the SIA were a decrease in the revenue of both the district municipality and the local municipality, the termination of service agreements with the mine, the closing down of the mine clinic and an increase in indigent households (CDS, 2005).

The SIA raised concerns about the land use and infrastructure impacts of mine closures. Recreational facilities would deteriorate. The Koffiefontein mine owned and maintained most sports, recreation and entertainment facilities (a form of interdependency: the employees depended on the mine for these extras, and the mine depended on having a contented workforce). The community and the local municipality had little obligation to maintain these facilities. There would be a decrease in property values and an increase in the number of vacant properties (Marais et al., 2005). Direct negative economic impacts would be the loss of job opportunities for employees, the downscaling of mine-related and mine-dependent businesses, accompanying job losses, and the loss of service benefits and allowances for mine employees. The SIA expected a loss of community identity, increased community isolation, increased individual and community deprivation, fatalism and negativity, and a decline in the general quality of life (CDS, 2005).

In sum, the SIA of 2005 provided the Koffiefontein mine with a foundation to manage the consequences of mine downscaling. Erzurumlu and Erzurumlu (2015) describe the negative consequences for communities of poorly managed mine closures. They call for a stakeholder-inclusive approach to planning boom-and-bust cycles. Such an inclusive approach should help to achieve a wider acceptance of closure outcomes.

9.4 Findings

Mining downscaling causes many problems for communities. In the following, we list some that we found in Koffiefontein.

9.4.1 Poor financial planning by retrenched miners

Ensuring that workers who lose their jobs make sensible financial decisions can help to mitigate financial distress. Ackermann et al. (2018) note that mine closure can do irreversible damage to miners' chances of being re-employed. The dependency lies in the nature of mine work: there are not many related occupations to which they can transfer their mining skills. They thus struggle to find employment in other economic sectors. More than two decades ago, De Beers (2009) emphasised the importance of supporting workers with financial planning and debt management. This is not easy when a mine closes. Our study showed how downscaling led to problems with household finances. Our interviewees said they were grateful for the employee wellness programmes but that a lack of income was causing them a great deal of stress, and retrenchments had led to the disintegration of their families. One interviewee said that being unemployed 'created unbearable tension in the family', and another was 'not surprised that many ex-mineworkers separated from their partners here'.

This situation has been exacerbated by the failure of retrenched miners to anticipate and plan for the downscaling (an example of interdependency). We saw that negligence and ignorance on the part of these miners had resulted in insecurities. James (not his real name), a 58-year-old man who had worked at the Koffiefontein mine for over 20 years, was a typical case. When the mine retrenched him, he used his severance package to buy two cars. Like most others in his situation, James lamented that 'Koffies [as Koffiefontein is affectionately called] was the place to be. We partied daily without any care about the future'. As time passed and the money dwindled, he sold one of his cars to get some cash. His wife left him, and he currently depends on friends and relatives. The 2005 SIA predicted that closure would cause this kind of problem (CDS, 2005). And this indeed seems to be the case.

9.4.2 Outmigration

Again, this was predicted by the SIA in 2005. As employment opportunities dwindled, those with skills left. A municipal official had this to say about the unintended consequences of mine closures:

> One that comes to mind is you lose a lot of skilled people. Skilled personnel, people with skills, would immediately leave. And these would be people who probably were professionals in the mine and immediately would have to go and look for work elsewhere.

This indicates that it is more than just a matter of the outmigration of unskilled labourers: Koffiefontein is losing human capital. The loss of skilled people is a major concern, as these people might have helped to create a post-mining economy. The 2022 Census data, unavailable at the time of writing, will give a better idea of the population trends, but most interviewees were aware that people were leaving Koffiefontein.

9.4.3 Closure of businesses

Mining played an essential role in Koffiefontein's economy through employment opportunities and the circulation of capital. The interviews show that the decreased buying power and outmigration caused by mine downscaling were bad for businesses. Siyongwana and Shabalala (2018) give a detailed account of the way that mine downscaling affects local enterprises. In Koffiefontein, one of the notable effects of mine downscaling was the closure of businesses, including banks. Business owners agreed that new enterprises were unlikely to be created in the absence of the mine. One of them said he had had to close and rent out one section of his premises to maximise income. Another said there was no single route towards economic growth, but he felt that the mine and the municipality should 'explore different processes for ensuring local economic prosperity'.

Small businesses that depended heavily on mining, such as those involving hawkers, taxis and spaza shops, became unprofitable. The Letsemeng IDP acknowledges the vulnerability of local businesses by calling for collaborative planning to create economic development (Letsemeng Local Municipality, 2020). But we saw little evidence of success. Business owners expressed concerns about how Koffiefontein has lost its status as a trading centre in Letsemeng. One of them said, 'When you wanted to do proper shopping, you would come to Koffiefontein, but, sadly, Jacobsdal [a nearby rural town] has become the economic hub of Letsemeng'. This decline in economic importance aligns with the literature on boom-and-bust cycles (e.g. Haikola & Anshelm, 2020; De Beers, 2009). The finding also confirmed predictions in the 2005 SIA.

9.4.4 Increase in sex work and teenage pregnancy

Interviewees told us they were aware of an increase in prostitution and teenage pregnancies, closely related to issues of family disintegration. The SIA predicted isolation and despondency. This could be partly responsible for the increase in these problems. One interviewee said those involved in particular were 'girl children and those who work but cannot make ends meet' and were thus obliged to 'use their bodies as a source of income'. Another said:

> Because there is no work, and they have needs and expectations, you find girls have a relationship with married people for money to survive. I've often heard mothers crying or complaining about the situation.

Sesele et al. (2021) found similar patterns of child abuse in the Free State Goldfields.

9.4.5 Increase in crime

Mine closures can increase crime. The overall logic is that when mine closure leads to unemployment, children drop out of school and resort to crime and drug abuse. The interviewees claimed that the town is not as safe as it used to be. Although the SIA did not predict an increase in crime, it did refer to several aspects of social disruption related to closure. The following comments (one from a retrenched miner and one from a municipal official) describe the effects:

> When the mine closes, it becomes a problem because, for example, in a household, if I am the father and breadwinner, and I lose my job, respect is gone in the house because I cannot cater for my family. The house order is no longer there because the father has lost respect. He cannot sustain or maintain the house. So, the children become vulnerable, and they start using substances. They start doing all the wrong things, and when there is no income, my children drop out, which is where crime comes in.
>
> When the mine closed, one would expect an increase in the crime rate. In what we have seen even in recent times, there have been attacks on private individuals. The situation that would have resulted from unemployment gave

> rise to the increase in crime. Not very long ago, there was an attack on the police station where on-duty police officers were murdered. So that shows you how crime has ravaged that small town.

Although crime levels are generally high in Koffiefontein, the official figures show no substantial increase since 2010 to confirm these views. However, the murder rate in the Letsemeng Local Municipality of 42 murders per 100,000 of the population between 2010 and 2021 was the fifth highest of all municipalities in the Free State. It is higher than the 33 per 100,000 in the Free State province overall. Figures from one year of information show that cases of assault intending to inflict grievous bodily harm increased by almost 61% from April 2021 to March 2022. These figures suggest social disruption linked to mine downscaling.

9.4.6 Loss of livelihoods and lack of socioeconomic development

The decline in mining in Koffiefontein has adversely affected livelihoods, particularly for those directly dependent on the mine. The community leaders said the decline in mining had hit them particularly hard because there had been no warning signs. One interviewee said closing the mine 'also meant closing the town'.

Concerning the takeover by Petra Diamonds, interviewees thought the new owners needed to be more helpful. Ten years after its takeover, the company's SLP only indicates how the company will align its socioeconomic contribution with the municipality's IDP (Petra Diamonds, 2017). Some community leaders believed that the municipality and not the mining companies, first De Beers and now Petra, should be blamed for the lack of socioeconomic development in the town. One of them described the situation as follows:

> I know many residents will blame De Beers for the poor living conditions in this town. But what happened to the community's appeal during several IDP consultations about a brickmaking and tile-making factory? Such small-scale industries can make a significant contribution to the high unemployment rate. After so many years of mining in this town, our municipality still needs to learn about the importance of planning for the future.

The potential dovetailing of IDPs and SLPs holds much promise. However, implementation is difficult. This interviewee also said the municipality has been unable to ensure the implementation of many projects. Implementation requires skilled officials who understand relationships with the private sector. This skill seems to be absent.

Another community leader, explaining the municipality's passive approach to mine downscaling, said:

> SLPs are about what the municipality can gain rather than stimulating sustainable economic growth. SLP-related projects became a means of survival for the municipality, so they will consult us during the initial phase of projects

> but leave us in the dark regarding progress. So, this poverty-stricken community has been on its own in terms of dealing with the dire consequences of mine downscaling.

The comments about being 'in the dark' and the community being 'on its own' point to a lack of collaborative planning and engagement with the community. In addition, an official from the municipality reported that the municipality has no closure plans and would only be able to implement such plans with financial support from provincial departments. The community leaders quoted earlier doubted the municipality's ability to ensure that the SLPs will be implemented and that they will support the post-mining economy.

9.4.7 Deterioration of infrastructure

The government and the mine developed the initial infrastructure in Koffiefontein for a particular time and place. Infrastructure development was tied to mining. The SIA identified land use and the related infrastructure impacts as one of five cross-cutting issues in 2005 (CDS 2005). The effects of mine downscaling are visible in the deterioration of recreational facilities (parks and public open spaces), social amenities, commercial buildings, municipal services, decreased property values and vacant properties.

Figure 9.2 shows the increase in residential land use from 1990 until 2020. Despite declining economic activity, informal settlements are growing. Marais

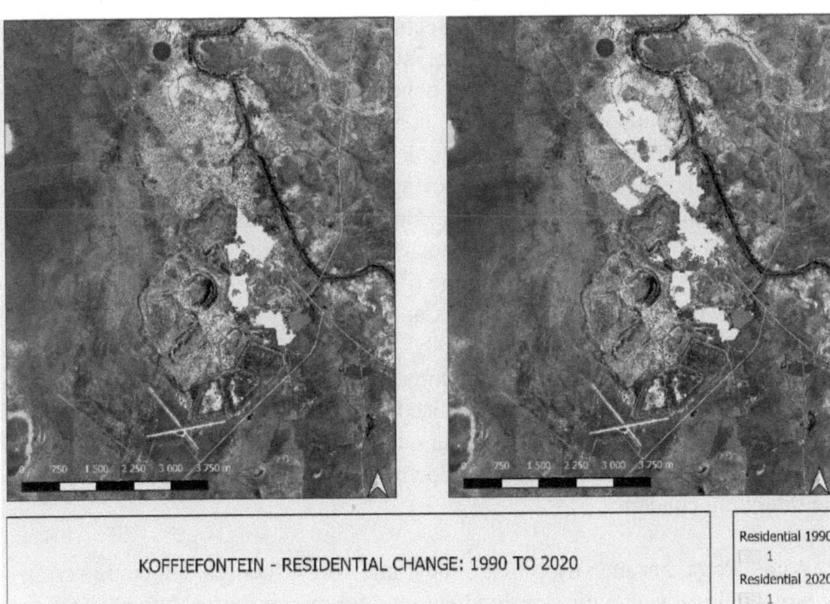

Figure 9.2 Residential land-use increase in Koffiefontein, 1990–2020

et al. (2020) note that the growth of informal settlements has been driven to a large extent by the South African government's ending the system of housing miners in compounds and discouraging labour migration. Another major impact of the normalisation of the housing system is the rapid increase in informal housing (Marais, Burger et al., 2022), leading to pressure on already resource-constrained municipalities' service delivery.

The increased land use has required increased municipal infrastructure. In a period of economic decline, maintaining infrastructure has become expensive. Deteriorating roads cause problems for business owners and residents. The interviewees were unhappy about the failure to maintain infrastructure and the collapse of the road infrastructure in Koffiefontein. They were concerned about the way the bridge over the Riet River near the town had been neglected. According to a business owner, the bridge has rapidly deteriorated and requires attention before a disaster occurs. Acknowledging the potential danger, the Free State Department of Police, Roads and Transport announced in early 2023 that it planned to revamp the bridge and the access road (News24, 2023).

According to a former municipal official, the municipality's capacity to maintain roads is under strain. He said, 'part of the explanation lies in municipalities concluding mining development agreements with immediate benefits in mind'. This emphasis on the 'immediate' contrasts with the required long-term thinking, including thinking about mine closure. Consequently, economic decline has become a reality in Koffiefontein.

The IDP says the municipality's sewer pipelines are ageing. Since installation, they have accumulated sand, debris and other foreign materials (Letsemeng Local Municipality, 2020). Originally, De Beers upgraded the local water purification plant and sewage system and provided the municipality with maintenance support – an example of the interdependencies that develop between mines and municipalities. But this type of support is unlikely to continue during downscaling or after closure. It did not continue when Petra took over. One community leader said 'the poor maintenance of wastewater and sewage treatment is a health risk to residents and the environment'. The deteriorating state of municipal wastewater and sewage treatment has affected the river basin ecosystem, causing pollution that has consequences for families who depend on the river for fishing.

De Beers donated several buildings and facilities to the municipality when it sold the mine to Petra Diamonds in 2007. While the donation could be viewed as generous, the municipal officials we interviewed noted that it had increased the municipality's asset base but compromised its income because it shifted the maintenance and rate liabilities from the mine to the municipality. The municipality is at the mercy of the mine's decision to downscale. Another municipal official raised the following concern:

> When Petra became owners of the mine, they focused on empowering women, local economic development, and funding infrastructure projects as identified by the community, like grading and gravelling of internal streets. But unfortunately the municipality is on its own regarding infrastructure

maintenance. And with the inadequate equitable share grants for the municipality, we cannot maintain facilities and infrastructure as we would like.

Unsurprisingly, business owners and municipal officials blame the mine downscaling on poor infrastructure management. The municipal officials believed that the Koffiefontein mine has provided only cosmetic infrastructure improvements to obtain community approval (an example of how the initial goal of obtaining the community's approval leads to the community expecting too much of the mine and coming to depend too heavily on it). Overall, there needs to be more evidence of collaborative planning between the mine and the municipality.

9.5 Discussion and conclusion

The chapter discussed the socioeconomic and infrastructural consequences of mine downscaling in Koffiefontein. Mine closure, while complex and inevitable, affects the livelihood of communities and municipal infrastructure. Mining almost invariably results in dependence, which creates complacency in decision-making at several levels. Planning is often path-dependent on growth related to an operational mine. Thinking about and planning decline receives little attention. But there are also interdependencies that, when broken, create problems. For example, the Koffiefontein mine not only maintained but also provided infrastructure. The mining operations necessitated this, but once the company began to downscale, it ceased to bother about it. The deterioration can be blamed on a failure to adapt to mine downscaling and imminent closure. Koffiefontein mine has been closed and reopened many times in the past. This should have provided the municipality and the community with a skill set to counter the damaging effects of downscaling.

Although mining legislation is explicit about the procedures to follow during mining decline or closure, mining companies seldom follow the regulations. This raises questions about the mines' responsibility to local communities. Our interviewees expressed mixed views about the Koffiefontein mine's responsibility. It also raises questions about whether the challenges faced by mining communities should be the sole responsibility of mining companies (Syahrir et al., 2021). Some lessons can be drawn from our Koffiefontein study, particularly in the case of the miners. For some who have now been retrenched, the instant gratification of an income blinded them to the need to plan for a reverse in the mine's fortunes.

Recognising the threat of mining to the environment and the well-being of communities, most mining-dependent countries now enforce regulations to reduce the damage of mine closures. Public-private partnerships have become pathways for infrastructure development, with the private sector providing design, construction, financing, maintenance and delivery. There have been some examples, but this has yet to find mainstream acceptance in South Africa.

Engagement with stakeholders has yet to be adopted as the core business of mining companies (Kemp & Owen, 2013). Communities and local authorities must reconsider their livelihood choices and governance decisions if they are to cushion the adverse socioeconomic and infrastructural impacts of mine downscaling. As legal sanctions will not remedy the way mine-community relations are managed because of historical concerns, a plausible alternative for mining communities is to begin to embrace practical solutions like a savings or investment culture.

Most of the 2005 SIA predictions have become realities. But neither the municipality nor the Koffiefontein mine have actively addressed these concerns. It is possible to predict the adverse consequences of mine closures. Predicting it also means that appropriate planning can take place. For example, knowing that infrastructure maintenance will be a problem should prompt municipalities to save in order to deal with this reality. At the very least, a municipality faced with mine closure should not take on new infrastructure or develop more land. Despite the consequences of closure, mining companies should be able to generate value for themselves and nearby communities, and this value could perhaps extend beyond mine closure.

References

Ackermann, M., Botha, D. & Van der Walt, G., 2018. Potential socio-economic consequences of mine closure. *Journal for Transdisciplinary Research in Southern Africa*, 14(1), pp. 1–11.

Atkinson, D., Schenk, R., Matebesi, Z., Badenhorst, K., Umejesi, I. & Pretorius, L., 2016. Impacts on the social fabric. In Scholes, R., Lochner, P., Schreiner, G., Snyman-Van der Walt, L. & De Jager, M. (eds.), *Shale Gas Development in the Central Karoo: A Scientific Assessment of the Opportunities and Risks*. Pretoria: CSIR (Council for Scientific and Industrial Research).

Bainton, N. & Holcombe, S., 2018. A critical review of the social aspects of mine closure. *Resources Policy*, 59, pp. 368–478.

Campbell, M., 2021. Perceptions of the impact of extractive industries on local government: The case of the Rustenburg mining town. In Marais, L., Campbell, M., Denoon-Stevens, S.P. & Van Rooyen, D. (eds.), *Mining and Community in the South African Platinum Belt: A Decade After Marikana*. Hauppauge, NY: Nova Science Publishers, pp. 119–140.

CDS (Centre for Development Support), 2005. *Koffiefontein Social Impact Assessment*. Bloemfontein: University of the Free State.

De Beers, 2009. *Report to society 2009: 4. Employees*. www.debeersgroup.com/~/media/Files/D/De-Beers-Group/documents/reports/library/rts09-employees-may-2010.PDF

Diamond Fields Advertiser, 1987. Koffie re-opening to cost R70-m over 2 years. 30 June, p. 11.

Erzurumlu, S.S. & Erzurumlu, Y.O., 2015. Sustainable mining development with community using design thinking and multi-criteria decision analysis. *Resources Policy*, 46, pp. 6–14.

Haikola, S. & Anshelm, J., 2020. Evolutionary governance in mining: Boom and bust in peripheral communities in Sweden. *Land Use Policy*, 93, p. 104056.

IISD (International Institute for Sustainable Development), 2021. *Global Review: Financial Assurance Governance for the Post-Mining Transition: Intergovernmental Forum on Mining, Minerals, Metals and Sustainable Development*. www.iisd.org/system/files/2021-09/financial-assurance-governance-for-post-mining-transition.pdf

Kemp, D., & Owen, J., 2013. Community relations and mining: Core to business but not "core business". *Resources Policy*, 38, pp. 523–521.

Letsemeng Local Municipality, 2020. *Letsemeng Draft Integrated Development Plan 2021/2022*. www.letsemeng.fs.gov.za/wp-content/uploads/2021/05/DRAFT-IDP-2021-2022.pdf

Marais, L., Burger, P., Campbell, M., Denoon-Stevens, S. & Van Rooyen, D., 2022. Is a just transition possible? In Marais, L., Burger, P., Campbell, M., Denoon-Stevens, S. & Van Rooyen, D. (eds.), *Coal and Energy in Emalahleni, South Africa: Is a Just Transition Possible?* Edinburgh: Edinburgh University Press, pp. 230–238.

Marais, L., Denoon-Stevens, S. & Cloete, J., 2020. Mining towns and urban sprawl in South Africa. *Land Use Policy*, 98, p. 103953.

Marais, L., Ndaguba, M., Mmbadi, E., Cloete, J. & Lenka, M., 2022. Mine closure, social disruption and crime in South Africa. *The Geographical Journal*, 188(3), pp. 383–400.

Marais, L., Pelser, A., Botes, L., Redelinghuys, N. & Benseler, A., 2005. Public finances, service delivery and mine closure in Koffiefontein (Free State, South Africa): From stepping stone to a stumbling block. *Town and Regional Planning*, 48, pp. 5–16.

McGill, D.C., 1991. *A History of Koffiefontein Town and Mine 1870 to 1902 and 1902 to 1980*. Kimberley: Northern Cape Printers.

Monosky, M. & Keeling, A., 2021. Planning for social and community-engaged closure: A comparison of mine closure plans from Canada's territorial and provincial North. *Journal of Environmental Management*, 277, p. 111324.

News24, 2023. Bridge is open. *News24*, 17 January. www.news24.com/news24/community-newspaper/express-news/bridge-is-open-20230117

Petra Diamonds, 2017. *Social and Labour Plan: Koffiefontein Diamond Mine*. www.petradiamonds.com/wp-content/uploads/KDM-SIGNED-SLP-DOCUMENT-ENGLISH-March-2021.pdf

RSA (Republic of South Africa), 1996. *The Constitution of the Republic of South Africa (Act No. 108 of)*. Pretoria: Government Gazette.

Sesele, K., Marais, L., Van Rooyen, D. & Cloete, J., 2021. Mine decline and women: Reflections from the Free State Goldfields. *The Extractive Industries and Society*, 8(1), pp. 211–219.

Siyongwana, P. & Shabalala, A., 2018. The socio-economic impacts of mine closure on local communities: Evidence from Mpumalanga Province in South Africa. *GeoJournal*, 84, pp. 367–380.

Stats SA (Statistics South Africa), 2011. *Census data*. Pretoria: Stats SA.

Syahrir, R., Wall, F. & Diallo, P., 2021. Coping with sudden mine closure: The importance of resilient communities and good governance. *The Extractive Industries and Society*, 8(4), p. 101009.

10 Alexkor and the Richtersveld community

Unlikely partners to mine diamonds together

Phia van der Watt

10.1 From fighting over land to mining diamonds together

In 2007, the Richtersveld community and the state-owned mining company Alexkor negotiated themselves into a pooling and sharing joint venture (PSJV). This stemmed from an arduous legal battle in which the Richtersvelders sought reparation for losses incurred since the 1920s when the state started mining diamonds on their land. On 14 October 2003, the Constitutional Court confirmed the community's right to claim and referred the case back to the Land Claims Court to determine the reparation (Constitutional Court of South Africa, 2003). The parties started negotiating, a gruelling process lasting another four years. On one side of the table were the Richtersveld community's Sida !hub[1] Communal Property Association (CPA) and their legal team from the Legal Resources Centre.[2] On the other side were Alexkor and the Department of Public Enterprises (DPEs), under its minister Alec Erwin.

On 26 October 2006, the parties signed a Memorandum of Understanding (DPE, 2006). The negotiations almost derailed in February 2007, when it emerged that the government was negotiating with another mining company, but on 22 April 2007, Erwin announced that the parties had signed a Deed of Settlement. This was lauded as a victory for the Richtersveld community and the finalisation of the land claim. But there were also grave misgivings about the secrecy surrounding the signing of the Settlement and some of the terms in this document. Some felt Erwin had coerced the leaders into signing and was using the reparations to the community to capitalise on Alexkor's failed operations (Cargill, 2010). The Legal Resources Centre, which had driven the Richtersveld claim process since the beginning and was intimately involved with the community, withdrew as the community's legal representatives. This became one of the first wedges driven into a community that had previously shown a united face throughout the land claim process (Cargill, 2010). The community ratified the Deed of Settlement on 4 August 2007 (Sida !hub CPA, 2007). On 9 October 2007, the Land Claims Court made the Deed of Settlement and the supplementary Unanimous Resolution an order of the court (Land Claims Court, 2007a, 2007b).

This Settlement tied the Richtersvelders into a joint venture with Alexkor and thus with the government, since Alexkor is a state-owned entity. Alexkor held a

DOI: 10.324/9781003403326-10

51% share and the Richtersvelders 49% in this Pooling and Sharing Joint Venture, or PSJV. Technically, most of the Settlement conditions had been met by 2012. Sida !hub established its comprehensive governance structure; the land, assets and mining rights were transferred to Sida !hub; the monetary grants were paid out and the PSJV was established with a joint board and executive management committee. The only outstanding matter was (and by 2023 still is) the transformation of Alexkor's mining village, Alexander Bay, into a normalised town under the Richtersveld Local Municipality.

However, neither party was in a good position when the PSJV was established. Alexkor was operating at a loss, and Sida !hub was locked in internal conflict. Matters did not improve, and by 2020, both Alexkor and Sida !hub were under administration. There was little leeway to solve problems emanating from the Settlement because it could only be changed through a new court order.

The chapter investigates the functioning of the PSJV and the partnership between the small and divided Sida !hub community and an almost bankrupt state-owned mine as the resolution of a land claim. It questions the appropriateness and potential of a high-risk settlement, where danger lights were flickering even before its signing: the bloated governance structure, partners mistrusting each other, and divisions within the community (specifically about the decision to settle). The mining literature often uses the term 'resource curse' to explain the mistrust and conflict when a mine opens. The evidence I present here shows that the curse continues even as mine declines.

To understand the community's perspective, I conducted interviews in the Richtersveld in July 2021. The 17 interviewees were mine contractors, business owners in Alexander Bay, a social worker, people involved with environmental rehabilitation, community members and three ex-leaders of the CPA (there was no legal CPA structure at the time of the interviews). I also talked to the administrator appointed to manage the affairs of the Richtersveld community from 2020 to 2023. The Alexkor Board declined a request to interview its managers or employees. I turned to discussions and presentations of the Portfolio Committee on Public Enterprises and Communication in parliament and to Alexkor's annual reports to learn about developments in Alexkor, the PSJV and, in some cases Sida !hub. I also accessed unpublished Sida !hub documents, including their constitution and the 2010 Development Plan (see Figure 10.1).

10.2 The PSJV as described in the Deed of Settlement

The Deed of Settlement is a complex legal document that combines the terms and outcomes of a land claim with detailed arrangements regarding the PSJV. It includes clauses on the transfer of assets to Sida !hub, as well as the governance and financial arrangements of the PSJV and contractual obligations. The following are the points most relevant to this chapter:

Sida !hub received approximately 84,000 hectares of diamond-rich land from Alexkor, Alexkor's agricultural and mariculture farms (with a R50 million recapitalisation grant), and almost all the properties in Alexander Bay. Alexkor could rent

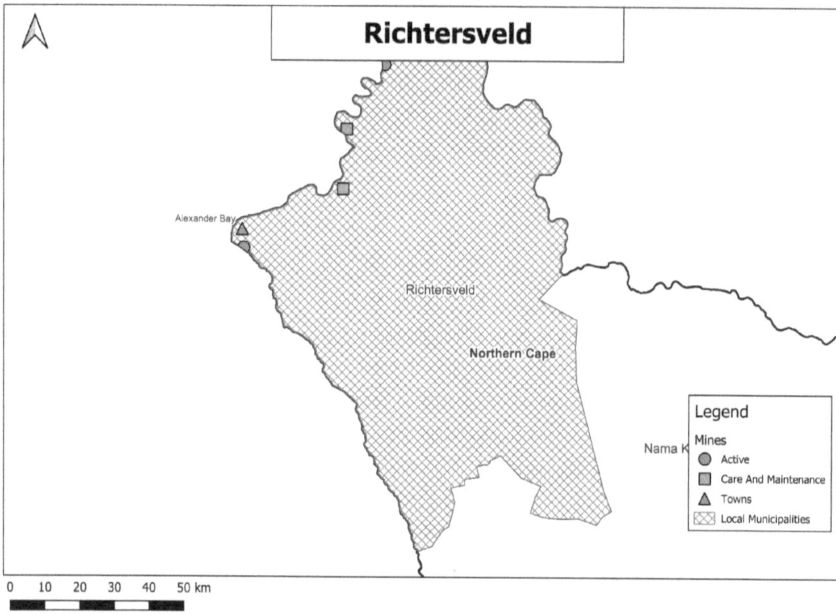

Figure 10.1 Mines and settlements in the Richtersveld

these for ten years from Sida !hub for R45 million. Sida !hub also received a R190 million compensation grant from the state, to be invested and used for direct payouts to members and for development in the Richtersveld.

The two parties would form the PSJV, in which Alexkor has a 51% interest and Sida !hub a 49% interest. Alexkor would retain its marine mining rights[3] but have to transfer its land mining rights to Sida !hub. All mining rights (marine and land) and assets were pooled in the PSJV and controlled by the PSJV and Joint Board, comprising three members each from Alexkor and Sida !hub. The three Sida !hub representatives were to be drawn from one of the entities it had to establish, namely the Richtersveld Mining Company (RMC). The Joint Board would appoint an executive committee responsible for the day-to-day operations of the PSJV (such as a CEO and financial and mining managers). A recapitalisation loan of R200 million was made available. All mining was to be done by contractors. The life-of-mine was estimated at an annual production rate of 150,000 carats for ten years. Alexkor and the state were responsible for all historical environmental damage until the day of transfer, whereafter the PSJV would be responsible for damage caused by its operations. Alexander Bay, Alexkor's mining village, had to be transformed into a normalised town under the Richtersveld Municipality. Before handover could take place, Alexkor had to upgrade services and infrastructure to meet municipal standards.

To receive, manage and protect the assets obtained through the Settlement, Sida !hub had to set up eight legal entities besides the 15-member Sida !hub CPA. This implied a governance structure of 74 trustees, directors and members, of which 40 were to be elected or appointed by the approximately 3,000 members of the Richtersveld community. These entities are the Investment Trust, the Investment Holding Company, the Self-Development Company or Self-Devco, the RMC, the Richtersveld Property Holding Company, the Richtersveld Rehabilitation Company and the Richtersveld Agricultural Company.

The Settlement provided for the dissolution of the PSJV. Alexkor had to withdraw from the PSJV through a call option (the right to buy) to be given to the RMC to buy Alexkor's 51% interest in the land mining operation. This could happen immediately after transferring the land mining rights to the RMC, but not later than five years after that. On the exercise of the call option, the PSJV would dissolve. Thereafter, Sida !hub and Alexkor would retain their respective mining rights, operations, resources and assets (LCC, 2007a, 2007b)

The land mining rights were transferred to Sida !hub on 6 April 2011, and the PSJV and the Joint Board were established on 7 April 2011. The Executive Committee was formed in 2012. Since the Deed of Settlement was made an order of a court, another court order would be required to change any aspect thereof.

10.3 The PSJV

> The old biographer's problem: even when people are telling their own life stories, they are invariably improving on the facts, rewriting their tales, or just plain making them up.
> (Salman Rushdie)

10.3.1 A shaky foundation for a complex business

Sida !hub and the DPE/Alexkor started negotiating the terms of reparation in (2003) and it took them four years to reach a deal. After fighting each other for a decade in the country's courts, there was a trust deficit. An interviewee who was part of the negotiations explained their dilemma: on the one hand, they wanted to get the best deal possible for Sida !hub; on the other, they feared that if they dragged the process out, they could be left empty-handed. It was clear that Alexkor was not taking proper care of the assets they would have to transfer to Sida !hub (such as the farms and the infrastructure of Alexander Bay) and, in addition, the mine was already running at a loss. The community was becoming impatient with the lack of progress. They knew they could not independently manage a mine, so they looked for a partner – anyone other than Alexkor. However, they were made to understand that certain legalities, like the transfer of mining rights to a community, could be challenging. Instead of risking further delays, the leadership opted for a bridging arrangement: a joint venture with Alexkor. After completing all transfers, they would be able to exit the partnership and seek a partner of their choice.[4]

With the advantage of hindsight, some interviewees rejected the aforementioned arguments, perceiving them as Erwin's strategies to force Sida !hub into a partnership that would be convenient for the DPE. The government was also anxious to reach a deal because Alexkor was an insoluble problem for the state: it had become a financial drain, but the state could not embark on strategic changes or get rid of it in the middle of a pending land claim (Joubert, 2007).

In October 2006, the parties signed a Memorandum of Understanding, providing for a PSJV similar to the one agreed on in the eventual Settlement (DPE, 2006). The negotiations were almost derailed at the beginning of 2007 (Market News, 2007), but in April 2007, the CPA leaders signed the Deed of Settlement with the DPE and Alexkor. There are many versions and interpretations of the events between April and October 2007. Even though the Settlement was essentially based on the widely accepted 2006 MOU, certain factions opposed it and described it as flawed. There were rumours that there had been suspicious dealings between the CPA leaders and Alec Erwin and that the leaders had agreed to clauses against the advice of their legal representatives, the Legal Resources Centre. The latter suggested to journalists that the CPA leaders might have been 'bribed, forced or bullied into signing' the Settlement (Joubert, 2007).

There is some truth in this assertion: Erwin often threatened the community publicly and warned them not to expect too much. A 2007 Sida !hub leader recalled that Erwin threatened to close Alexkor and sell its assets if the community did not agree to a settlement, which would have implied the loss of 800 jobs (mostly those of Richtersvelders). The leaders believed that it was time to settle before they lost too much. They feared that there would be nothing left of Alexkor's farms[5] and the mine. And there remained the critical question: what was the alternative? Should they have continued with legal action? For how many more years? Who could guarantee the result of that route?

The signing of the Settlement was the most significant event in the Richtersveld land claim – not only because it heralded the conclusion and tore the community apart but also because it forced them into a complex joint venture with Alexkor. The foundation on which the PSJV was built was indeed rather shaky.

10.3.2 Governance through absence, duplication and conflation

The governance of joint ventures involves at least three structures: those of the two (or more) participating partners and the governance structure of the newly established venture. In this case, the structures of the two participating partners are Sida !hub's CPA structure, with all its entities, and Alexkor with its board. Both structures would continue to exist beyond the PSJV. The third structure is that of the PSJV, the Joint Board, which was supposed to have a limited life (as is the case with all joint ventures).

10.3.2.1 The Sida !hub CPA

For most of the years between 2010 and 2022, Sida !hub did not manage to set up or maintain a CPA committee or the two trusts and six companies required by the Settlement. Meetings, elections and decisions were constantly disrupted, disputed or interdicted by an opposing faction. The main reason has been internal conflict, which was the topic of various investigations and interventions by government departments and the Portfolio Committee on Public Enterprises and Communication. The Sida !hub Development Plan (Sida !hub CPA, 2010) dealt extensively with this issue. It highlighted problems with voting procedures, jealousy between the villages and factions, nepotism and the community's attitude towards authority and outsiders.

Most interviewees made it clear that they were tired and ashamed of the quibbles, the undermining of their leaders and the factions that paralysed them all. There was always a group ready to embark on new litigation. An interviewee was so desperate that she suggested they stop holding elections altogether and instead appoint an expert from outside to run the community's affairs. Others suggested setting criteria for nominations and deliberately bringing in younger leaders to stop old feuds from being brought into every decision. There were also signs of the deliberate spreading of rumours and disinformation. Wild allegations were thrown around, with one interviewee alleging that a leader 'stole millions of rands'. When challenged to provide detail, the interviewee wanted to stop the interview. A story was spread that the community should insist on the immediate pay-out of the R45 million rent for Alexander Bay's property. This is legally impossible, but the idea spread like wildfire and became a new obsession.

All interviewees suggested that Alexkor's meddling in their affairs fuels the conflict. They believed Alexkor sought to keep the entity dysfunctional so that it could run the PSJV as it wished. Interviewees accused government officials and politicians of inappropriate interference because they wanted to benefit from Sida !hub's diamonds and other regional developments (such as the proposed multi-billion-rand Boegoebaai Harbour and green hydrogen projects).

The department responsible for land reform drafted numerous reports on conflict within Sida !hub. Their findings are supported by data obtained from the interviews: relentless infighting, ongoing litigation, an unwillingness of committee members to vacate their positions, Alexkor's meddling in the community's affairs, and consultants becoming part of the conflict. The reports also noted that the community does not understand, or wants to understand, the binding directives of the Settlement (the R45 million issue was used as an example). The efforts of the departments of public enterprises and land reform had no effect (CRLR, 2018; DALRRD, 2022; Fife, 2013; PMG, 2019a).

Another reason for the conflict in the Sida !hub is the legal system of CPAs, which was designed to facilitate land claims. During a discussion in the Portfolio Committee on Public Enterprises, officials of the department for land reform spoke about the system as 'by its very nature . . . conflict-ridden'. When challenged to explain why it was not transforming or scrapping an instrument that the department itself recognised as being flawed since inception, officials said they were still investigating alternative models for land-holding entities (PMG, 2022).

The implosion of its governance system had devastating consequences for Sida !hub. No money can be paid out to members without an approved membership list. This approval is only possible through elections and annual general meetings, which have been disrupted or interdicted year after year. Sida !hub thus cannot establish the entities required by the Settlement to receive the assets and money it was granted, and, critically, it cannot be represented in the PSJV and Joint Board because there is no legally constituted RMC. As a result, the court appointed independent directors on behalf of Sida !hub. In effect, Alexkor runs the PSJV alone.

10.3.2.2 Alexkor

Through the Settlement, Alexkor lost its land mining rights to the Sida !hub, and even though it retained its marine rights, these were pooled into the PSJV. Alexkor's governance system (its board) remained untouched after the Settlement, as is expected when a company enters into a joint venture. However, almost all non-core activities (e.g. the farms and the airport) were transferred to Sida !hub and state entities. Technically, Alexkor was left with two tasks: to upgrade Alexander Bay for transfer to the Richtersveld Municipality and to participate (through its board representatives) in the PSJV. It has nevertheless maintained a high-level management team. It was a mystery to interviewees what Alexkor's mine managers do since all mining activities are now in the hands of the PSJV's management team, which includes a CEO and financial and mining managers. And interviewees found it annoying that these managers are based in Sandton and fly at great cost to Alexander Bay for meetings. This issue is indeed puzzling: Alexkor has been in a dire financial state since the early 2000s but continues to pay highly qualified mine managers without a mine to manage.

Since 2007, much attention has been paid to Sida !hub's bloated structures, while Alexkor's governance structures have escaped similar scrutiny. This changed in 2017 when the duplication of the governance and management systems of Alexkor and the PSJV came under the spotlight. A waste of at least R4 million per month was discovered. Steps were thus proposed to avoid duplication. The following options were under consideration at the time of writing: collapse the two boards and administrative structures; close the Johannesburg office; transfer Alexkor's marine mineral rights to another company; transfer the remaining Settlement obligations to the DPE and move the Alexkor CEO to Alexander Bay (PMG, 2017, 2020a, 2020b, 2020c, 2022).[6]

Almost all the aforementioned steps require legal scrutiny. Sida !hub is constantly warned that they cannot change a letter of the Settlement without a new court order. Would collapsing Alexkor and PSJV structures or transferring Alexkor's marine mining rights not violate the Settlement? What would the relationship be between the PSJV CEO and Alexkor's CEO if the latter was moved to Alexander Bay?

Since Alexkor's mandate is the mining of diamonds (and this was not exactly profitable within the PSJV), the DPE and Alexkor decided to diversify into coal and limestone mining, ostensibly to make Alexkor more sustainable. However, nothing eventually came of this plan because the scope and modus operandi of state capture[7] started to surface before the plan could be implemented. Some damage had been done, though: since 2012, various ministers of public enterprises have started to appoint people linked to state capture to strategic positions on the Alexkor Board and as managers of Alexkor and the PSJV. These directors and managers awarded a tender for the sale of the PSJV diamonds to Scarlet Sky Investment, a company without a diamond trading licence but with links to state capture (Commission of Inquiry into

State Capture, 2022). Contractor interviewees believed they had lost money in this period (2014–2016) when their diamonds were sold at below market value.

10.3.2.3 The elusive business of the PSJV

A 2017 DPE report complained about the 'lack of visibility into the affairs of PSJV' (PMG, 2017). I could also not find documents explaining the business of the PSJV as a separate entity (such as annual reports, statistics or audits), but I did find information about the PSJV in Alexkor's annual reports. The 2016 annual report notes that Alexkor was mandated to oversee reporting by the PSJV, as the PSJV operated as a division of Alexkor for reporting purposes. It does not explain how this mandate was obtained, as there is no reference to this in the Settlement. The information about the PSJV is dealt with separately in the earlier annual reports but becomes increasingly conflated in the later ones. An example is a reference in one report about the R30 million that Alexkor has spent on remuneration packages in the villages. This prompts the question: who is Alexkor employing in the villages, and for what work? In 2018, the term 'PSJV' disappeared almost entirely from the Alexkor reports. It was striking during interviews that the terms 'Alexkor' and 'PSJV' were used interchangeably. It is thus almost impossible to create a clear picture of the PSJV's governance or performance (Alexkor, 2012, 2020).

10.3.3 Contractors: mining with wheelbarrows and tied hands

The PSJV does not mine through employees but through contractors. Most of the interviewees had tried their luck as mining contractors with the PSJV or other mines in the area. They explained the system: the mine advertises contracts, which specify the area, types of mining (e.g. bedrock sweeping), what machines are allowed, the period of the contract and the split (e.g. 80% for the contractor). The contractors then have to find start-up capital (between R1 million and R15 million, depending on the contract), and this is almost impossible because they cannot offer guarantees to investors. Thereafter, it is a matter of luck because one could become rich or go bankrupt.

Contractors bring their gravel to the sorting plant, which is managed by the PSJV and where the gravel is processed. When diamonds are found, they are put in a bag, the contractor's name and the number of carats are noted, and the bag is sent to Johannesburg to be sold. Before contractors get their money, costs are subtracted. (For example, for security at the plant and 1% for 'the community'. None of the interviewees had ever heard of money that came to them from this 1%. They also cannot guess what else happens to it.) The process is identical for marine contractors.

Interviewees believed it was impossible to succeed as a contractor. One said, 'All you can do is to hope that you will find somebody who can bless you' (a South African term for someone who supports you financially; an investor or financier). The small contractors often get the lower-grade blocks and are limited to a spade, pickaxe and wheelbarrow. It is 'impossible to get anywhere with your hands tied behind your back', as one of them said. What causes bitterness is that when the small contractors give up, the big contractors are brought in without limitations on equipment. One interviewee was a supervisor at a new project, where a group of about 40 Richtersvelders got bedrock contracts. Many restrictive conditions forced them to go there with only wheelbarrows and a few workers. In addition, they had to pay salaries and about R1,000 per worker for health clearance, and they also needed a vehicle to get the gravel to the sorting plant. Some went to financiers, to whom they promised 'the sun, the moon and the stars'. And then, if they were not lucky in the first weeks, they started to accumulate debts and if the financier wanted his money back immediately, 'he drags you to court because the product you promised is a handful of gravel – not diamonds'. They were then worse off, with debt, an unhappy investor and unpaid workers. Within three months, there was nobody left. The interviewee said, 'It was very emotional to see how these people's hopes were dashed and how they crashed, one after the other'.

Something that irked interviewees was Alexkor and the PSJV's misrepresentation of the success of the contractors. They would declare excellent statistics in parliament just to look good – without mentioning that the positive numbers referred mostly to non-Richtersveld contractors. To counter this, a group of 17 Richtersveld contractors, including one of the interviewees, established the Richtersveld Contractors Forum in 2017, aiming to expose this misinformation in parliament. They stated their case, but nothing subsequently changed.

One interviewee reckoned that 'the contractor system is one of the saddest things that has ever happened here'. Another confirmed that you can expect only misery from a contract: 'I am not even trying to get a contract, because we learn from others' mistakes'. Instead, he said, you just go and find a small diamond or two near the river and enjoy life for a while: 'They know we do this and we see their helicopters looking for us, but they cannot trap us'. Ironically, when asked about the problem of illegal mining, the same interviewees told me that there was no illegal mining in the Richtersveld; this is a problem that only exists south of Port Nolloth.

Two problems with the PSJV's sorting and screening plants were reported. The first was the long waiting period, forcing contractors to wait weeks for their gravel to be processed. The second was questionable security at the plants, an issue raised by all interviewees as well as during discussions in the portfolio committee in parliament. The PSJV management consistently refused to bring in polygraph tests because, according to most interviewees, they knew exactly who stole – because they had appointed these workers (to steal).

The 2022 interim Alexkor board corroborates the interviewees' concerns about the small contractor system: it was not feasible since they could not secure financing. In addition, the PSJV could never reach its targets through this system. The board considered using only big contractors, who, in turn, could employ and empower small contractors. They also considered directly employing mineworkers again to generate the necessary revenue.[8]

10.3.4 Mining without maps

Prospecting is critical for a contractor; there must be some indication of how many diamonds are left in the block you intend to mine before you apply for a contract or seek an investor. Without reliable prospecting data, a contractor simply gambles when trying to get a contract. According to all interviewees, no prospecting work is done by the PSJV. Contractors thus work with inaccurate historical data. Alexkor's CEO mentioned the loss of geological information when the PSJV was formed,[9] resulting in a loss of R63 million (PMG, 2019b). One interviewee had worked on a project in Boegoebaai. The site manager had a map indicating a very rich, virgin area. There was also another map from the previous dispensation, which showed that the area had been mined. After massive input, the old map proved to be correct. The contractor he worked for lost a lot of money.

The Alexkor interim board believes that the appointment of bigger contractors could also help to solve the prospecting problem because they have the capacity to do exploration; the PSJV and small contractors cannot do it. They further identified an urgent need to recapitalise the mining operations, which can only happen if the residual deposits are confirmed. This will require costly exploration projects (Alexkor, 2020).[10]

10.3.5 Closing holes and turning pages: wiping out environmental footprints

The PSJV shifted the responsibility for rehabilitation to the contractors. An interviewee with a contract said, 'We are supposed to work like a mole, closing each hole we made; turning each page'. This is not an easy way to work because you lose track of where you have already mined and where you have not. The interviewee wondered what one would see on exploration maps over ten years. There is an environmental inspector at the PSJV, but interviewees were sceptical about whether she even goes to the mining sites. This fractured approach to rehabilitation is bad news for the environment. The damage is especially serious at the beach, where illegal coffer dams are being built.

10.3.6 Sharing the profit between the partners

The Settlement stipulates that the profit of the PSJV has to be shared between Alexkor and the community according to their 51%/49% share. No interviewee could enlighten me on when and how the sharing happens (or is supposed to happen). The Settlement does not refer to the contracts between the PSJV and contractors, which have a critical impact on the 51/49 split. If the contractors take 70% of the profit from the diamonds, is the remaining 30% all that is shared between Sida !hub and Alexkor? Should the contractor also share some of the profit with the community? Some interviewees pointed out that investigating this issue is a waste of time because there has not been any significant profit to share for years. It thus seems that the only Sida !hub members who benefit from the PSJV are those who are lucky enough to find a few diamonds as contractors. Some interviewees referred to a bank account where shared profits from the PSJV were deposited, but since there was no legal Sida !hub entity in place at the time of the research, I could not explore this further.

The 2022 interim Alexkor board also concluded that the Settlement did not specify how the sharing of profits and debts in the PSJV should happen and that it made no provision for the additional layer of joint ventures (PSJV contractors). The interim board was concerned that the community was not benefiting due to the contractor system.[11]

10.3.7 The PSJV's performance and turnaround strategies

The PSJV did not perform well as a business venture. Carat production dropped from 227,000 in 1998 to 43,000 in 2006 and 28,246 in 2021. It only increased to 33,066 carats when the Muisvlak plant opened in 2014–2015 and when 67,724 carats were found in 2019. There were profits only in 2012–2014 and in 2016/2017. The year-on-year financial performance is declining. In 2019, 150 employees were retrenched, and in 2020 another 80 (Alexkor, 2012, 2020).

By 2019, the department responsible for land reform had concluded that the Settlement was unimplementable and required a revision. The judicial administrator, appointed in February 2020, reviewed the Deed of Settlement and all the legal entities (DALRRD, 2022; PMG, 2019a).

The DPE appointed an interim board for Alexkor in February 2022, and they found Alexkor's governance in a 'dire state': basic governance and accountability measures were absent. Audits and reports for the previous year were not prepared, and no corporate plan or budget could be found. Since 2017, Alexkor has not received a clean audit. The interim board was shocked to find a total absence of role separation. The same person was simultaneously the acting CEO and director of Alexkor and the PSJV's CEO and board chairman. Measures taken by the interim Alexkor board and CEO[12] started to yield results, and by the end of 2022, production was slightly up and a small profit was expected for the 2022/2023 financial year (PMG, 2022).[13] Thanks to the administrator's work, things improved at Sida !hub and elections and an AGM were held in 2021. However, in 2022, the department reported that a new conflict had arisen within the committee, and some elected members had appointed attorneys to intervene in most matters.[14]

10.4 The impact of the land claim and the PSJV on the community

Retrenchments by the PSJV and at other mines in the Richtersveld had a devastating economic impact. Some skilled workers found work at the new Gamsberg mine near Aggeneys, but this meant that families were broken up. Ex-Alexkor and PSJV workers wanted to stay in Alexander Bay to access schools and healthcare and be as close as possible to work opportunities. They simply remained in their old houses, but as these houses belonged to the mine, they feared eviction. Alexkor does not maintain these non-employees' houses. The social worker I interviewed said some families had eventually left Alexander Bay and moved in with family members all over Namaqualand, which caused new problems for all the families involved.

Physical evidence of the retrenchments and the struggling mines can be seen in the obvious collapse of economic activities in Alexander Bay. Shops have closed, and finding even the most basic necessities is difficult. Interviewees complained that Alexander Bay looked uglier by the day because nobody cared since it was supposed to be transferred to the Richtersveld Municipality, but this had not happened. Alexkor is either unwilling or financially unable to maintain the town that it was supposed to upgrade for the transfer, and the municipality will not accept the town in its current state.

The effects of the resource curse can be seen in every corner of the Richtersveld. The social disruption caused by expectations of wealth and the subsequent disappointment is evident. Hungry children beg in the streets, and interviewees mentioned increased alcohol and drug abuse and crime. Interviewees from Alexander Bay told me that if you go away even for a few days and there is no light in your house, you must be prepared to buy a new TV and electrical equipment; there will be a burglary. In Sanddrift, interviewees were concerned about families sending their daughters to sell themselves to mineworkers for money or food.

Perhaps the most devastating result of the land claim is at the psychological level. The social worker who had worked in this community for decades described an atmosphere of depression: people have simply given up. When they won their land claim, they expected immediate wealth and they continued to hope for many years that this would be realised. The reality they now face is that they have gained nothing, apart from two insultingly small pay-outs: R1,500 in 2008 and R3,000 in 2011 (Fife, 2013). They cannot even access their own money because they have not managed to elect the necessary governance structures; as one interviewee put it, 'Our food stores are full, but we cannot get into them'. Some are angry, others ashamed; they are fools in the eyes of the whole country because they could not even make a successful land claim.

Shortly after the Settlement was signed, people were already fighting, but they were passionate then, and there was an expectation that something would change. By 2021, there was simply no energy left. An interviewee attributed the spirit of despondency and hopelessness to the lack of leadership. Nobody speaks or lobbies. Nobody is in charge, and one respondent said, 'The sting is out of our community'. Everybody sits and waits. Except, of course, the small groups plotting new interdicts and court applications.

10.5 Could the PSJV have succeeded?

Almohsen (2015) defines a joint venture as a business enterprise where companies voluntarily enter into a temporary partnership. Critical dimensions of successful joint ventures include good governance, equal partners and effective planning for an exit strategy. Good governance is linked to well-defined and separated powers, functions and responsibilities, both internally and across the partners' executive committees and boards. The complexity and risks involved are clear when we consider the 40% to 70% failure rate of joint ventures (Almohsen, 2015; Farrell, 2014; Footanstey, 2022).

When assessed against these critical dimensions, it is clear that the PSJV never stood a chance of succeeding. The partners did not enter voluntarily into the venture; they never trusted each other, the governance systems were under pressure, there was no preparation for closure, and the partners were not equal. Since its inception, the PSJV has been vulnerable to mismanagement, poor strategic decisions and lack of accountability. In addition, it was guided by the rigid Settlement framework, which in turn was informed by a range of state agendas. An example is the complex and bloated governance system. The aim may have been to address the government's fear that land claim funds might be abused. But it did not take into account the reality that this small and closed community would battle to elect 40 people with the necessary expertise or how difficult it would be to avoid accusations of nepotism. Another mistake was to believe that mining through local contractors leads to empowerment. That system is not only unfair to contractors, but it also compromises the profitability of the mine.

Diamond mining in the Richtersveld has declined for more than ten years since the inception of the PSJV. The situation improved slightly during 2022 when the interim Alexkor board took over and Sida !hub managed to conduct elections. But the damage done by the land claim to relationships and the psyche of the Richtersveld community may be too severe to repair. The community has lived for too long in the expectation of permanent riches and the belief that they are exceptional – delusions which were fuelled by those who assisted them in the land claim battle. The eventual adverse results have destroyed trust in outsiders' efforts to support them, as well as in their leaders. It will be very difficult for them to trust anybody again. Even more devastatingly, it appears from the interviews that the sense of failure has been internalised.

10.6 Conclusion: blending a land claim with a declining state-owned mine

Mines close for various reasons: the depletion of the resource, changing demand, rising production costs, national or geopolitical events, new policies or legislation, poor management and corruption. In the case of the Richtersveld diamond mine, all these factors played a role, but one critical factor was added: a land claim. After losing the legal battle against the Richtersveld claim, the government circumvented direct monetary compensation through a business deal with the community. And this joint venture had to solve another problem for the state: it had to keep the near-bankrupt Alexkor afloat, although the company was losing its land mining operation and most of its infrastructure through the claim. The Settlement was an opportunity for Alexkor to effectively re-gain access to the assets it had to transfer to Sida !hub. This seemed like a good deal for the government. When signing the Deed of Settlement, the Richtersvelders knew very well that Alexkor was in a poor state. However, their options were limited: they had to choose between continuing with a legal fight (and hope) for a better deal and entering as soon as possible into a controversial and poor deal with a partner they did not want, before there would be nothing left to fight for. These are not good reasons for establishing a complex business partnership; on the contrary, the PSJV lacked every dimension critical for a joint venture to succeed.

Over the past 15 years, the deal has not succeeded in serving any of the aforementioned purposes. First, the claim ripped the Richtersveld community apart, and members are now so absorbed in their conflicts that they are unable to elect leaders, which is necessary for them to access their own assets and to participate meaningfully in the PSJV. Second, the Settlement followed the global trend to mine through contractors instead of through direct employment. This arrangement resulted in jobs being lost, which were not redeemed by the contractor system. The system merely resulted in frustrated and bankrupt contractors who could not reach targets set for production and profit. In addition, through the contractor system, an additional level of 'sharing' was created (contractor/PSJV and Alexkor/Sida !hub), which was not foreseen in the Settlement and undermines the notion of pooling and sharing. Third, those who negotiated the Settlement did not take into account a danger that was only given a name years later: state capture. The PSJV did not save Alexkor, and it did not give the Richtersvelders access to their diamonds. Did those who conceived this plan realise that a failed PSJV would translate into a failed land claim? It is clear from the story of the Richtersveld claim that the mining industry is too volatile to act as a mechanism for historical compensation.

There were some positive signs by the end of 2022, but the impact of the past two decades on the psyche, socio-economic circumstances and relationships within the Richtersveld community cannot be turned around by new management structures or a slight increase in production or profit.

Over generations, the Richtersveld community and Alexkor had been neighbours and depended on each other as employers, employees and neighbours. Despite their interdependencies, the relationship between them has essentially been built on mistrust, resentment about dispossession and a sense of being wronged. Since 2007, they have been locked in additional interwoven forms of dependency. A court order made them business partners, which in turn bound them together into the state's ideological agendas regarding state-owned enterprises and land reform. They are locked in this toxic interdependency because both can only survive and reap benefits from the claim and the mine if they stay together. In addition, the community remains trapped in a yearning for historical justice and dreams of the ever-elusive wealth from their diamond-rich land.

Notes

1 'Our land' in the Nama language.

2 The Legal Resources Centre is an organisation advancing the transformation of South Africa as a democratic society, using the law as an instrument. See https://lrc.org.za/.

3 Alexkor traditionally mined on land and in the sea. The marine mining rights remained in the hands of Alexkor and the land mining rights were transferred to Sida !hub. However, both types of mining used the same offices and infrastructure on the land that since 2007 belonged to Sida !hub (e.g. the gravel from the land and the gravel from the sea bed are screened at the same plants in Alexander Bay). Alexkor thus needed access to this infrastructure. This alone illustrates the need for some type of cooperation and the interdependency between them. Alexkor's marine mining could hardly function without Sida !hub's assets. Without Alexkor's marine rights, there would be no need for and nothing to pool into the PSJV.

4 I could find only two references to the dissolution of the PSJV: one in an Alexkor (2015) annual report and one in a parliamentary report (PMG, 2021). Both noted that the exit

was to be done in a socially responsible way. The Alexkor report further stated that it was the Shareholder's prerogative to execute the option. That statement does not seem to reflect the relevant clauses in the Settlement, though. An interviewee who was involved in the negotiations expressed frustration that the PSJV did not dissolve as promised. This was an important issue during the negotiations: the leaders were comforted by the promise of the PSJV as a transitional measure, after which they would be able to choose their own partner. This never materialised and has hardly been mentioned since.

5 Because the area is so isolated, Alexkor established farms along the Orange River to provide food for the workers and their families. These became part of Sida !hub's land claim. During the almost ten years of the claim battle, Alexkor started to neglect the farms (and other infrastructure and assets in Alexander Bay). The Richtersveld community's fear that by the time they received the farms, there would be nothing left for them was a big issue during the negotiations and contributed to the leaders' decision to enter into what was an obviously poor deal.

6 See also Portfolio Committee on Public Enterprises, 30 November 2022: www.youtube.com/watch?v=qMVn7dORM-A (No written version was available.)

7 'State capture' is the manipulation of policy formation and implementation by members of the ruling elite and associates for personal gain. An example in South African is the appointment of allies in influential positions in state entities (such as Alexkor) and in departments. The special relationship between the Gupta family and the president of the country (Jacob Zuma) led to the appointment of a commission to investigate allegations of state capture – and Alexkor was mentioned in this regard.

8 Portfolio Committee on Public Enterprises, 24 August 2022: www.youtube.com/watch?v=Tg4L3VHyduY (No written version was available.)

9 According to an interviewee, this was done deliberately during the court case to paint a false picture of the volume and value of the remaining diamonds.

10 Portfolio Committee on Public Enterprises, 24 August 2022: www.youtube.com/watch?v=Tg4L3VHyduY

11 Portfolio Committee of Public Enterprises, 30 November 2022: www.youtube.com/watch?v=qMVn7dORM-A

12 During all introductions of the 'interim CEO', it was never mentioned whether he was the interim CEO of Alexkor or of the PSJV. He appeared to be reporting on what was happening in the PSJV (e.g. issues regarding contractors and security, which seems to be PSJV-related). But the impression given was that he was actually part of the Alexkor Board – another example of the possible conflation of the two structures.

13 Portfolio Committee on Public Enterprises, 30 November 2022: www.youtube.com/watch?v=qMVn7dORM-A; Portfolio Committee on Public Enterprises, 24 August 2022: www.youtube.com/watch?v=Tg4L3VHyduY

14 Portfolio Committee on Public Enterprises, 30 November 2022: www.youtube.com/watch?v=qMVn7dORM-A

References

Alexkor, 2012. *Annual Report*. www.alexkor.co.za

Alexkor, 2015. *Annual Report*. www.alexkor.co.za

Alexkor, 2020. *Annual Report*. www.alexkor.co.za

Almohsen, A.S., 2015. Joint ventures in the oil and gas industry. Unpublished doctoral thesis, University of Calgary, Calgary.

Cargill, J., 2010. *Trick or Treat: Rethinking Black Economic Empowerment*. Auckland Park: Jacana.

Commission of Inquiry into State Capture, 2022. Report: Part IV, Volume 1. www.statecapture.org.za/site/files/announcements/654/State_Capture_Commission_Report_Part_IV_Vol_I.pdf

Constitutional Court of South Africa, 2003. Case CCT 19/03. *In the case of Alexkor Limited and the Government of the Republic of South Africa versus The Richtersveld Community and Others*, 14 October. https://collections.concourt.org.za/bitstream/handle/20.500.12144/2176/Full%20judgment%20(445%20Kb)-17491.pdf?sequence=1&isAllowed=y.

CRLR (Commission on Restitution of Land Rights), 2018. *Status Report Richtersveld.* https://pmg.org.za/committee-meeting/27996/

DALRRD (Department of Agriculture, Land Reform and Rural Development), 2022. *Presentation to the Select Committee on Public Enterprises and Communication*, 8 June. https://pmg.org.za/committee-meeting/35181/?via=homepage-feature-card.

DPE (Department of Public Enterprises), 2006. *Statement by DPE: A. Erwin and Richtersveld community sign Memorandum of Understanding*, 26 October. www.gov.za/erwin-and-richtersveld-community-sign-memorandum-understanding

Farrell, P.E., 2014. The 7 deadly sins of joint ventures. *Entrepreneur*, 2 September. www.entrepreneur.com/article/236987

Fife, T., 2013. Richtersveld restitution implementation challenges. Land divided: Land and South African Society in 2013, in comparative perspective. Paper presented at conference "Land Divided: Land and South African Society in 2013, in Comparative Perspective", University of Cape Town, 24–27 March.

Footanstey, 2022. *Joint Ventures: What are the Key Governance Issues?* 19 February. www.footanstey.com/our-insights/articles-news/joint-ventures-what-are-the-key-governance-issues/

Joubert, P., 2007. Richtersvelders split over deal with Erwin. *Mail & Guardian*, 20 September.

LCC (Land Claims Court of South Africa), 2007a. *Deed of Settlement: Land Claims Court of South Africa 151/98. In the matter between the Richtersveld Community and Alexkor Limited and the Government of the Republic of South Africa*, 9 October.

LCC (Land Claims Court of South Africa), 2007b. *Unanimous Resolution of the Interim Joint Board appointed in terms of clause 8.4 of the Deed of Settlement*, 22 April. Signed 31 August.

Market News, 2007. Dispute over De Beers' SW African diamond deal. *Market News*, 28 February.

PMG (Parliamentary Monitoring Group), 2017. Report of the Portfolio Committee on Public Enterprises on Budget Vote 9: Public Enterprises, and the Annual Performance Plan for 2018/19 of the Department of Public Enterprises, 10 May. https://pmg.org.za/tabled-committee-report/3333/

PMG (Parliamentary Monitoring Group), 2019a. Presentation by the Department of Rural Development and Land Reform to the select committee on Communications and Public Enterprises, 27 February. https://pmg.org.za/committee-meeting/27996/

PMG (Parliamentary Monitoring Group), 2019b. Alexkor progress report on Deed of Settlement obligations. Report to NCOP Public Enterprises and Communication Committee, 6 November. https://pmg.org.za/committee-meeting/29268/

PMG (Parliamentary Monitoring Group), 2020a. State-owned companies update; DPE 2019/20 Quarter 3 performance, *Public Enterprises*, 26 February. https://pmg.org.za/committee-meeting/29899/

PMG (Parliamentary Monitoring Group), 2020b. *Alexkor Administrator on Outcomes of Forensic Investigation and Turnaround Plan*, 27 May. https://pmg.org.za/committee-meeting/30317/

PMG (Parliamentary Monitoring Group), 2020c. *Presentation by DPE on: Forensic Investigation at Alexkor and Progress Made in Implementing Turnaround Strategy to Portfolio Committee*, 11 November. https://pmg.org.za/committee-meeting/31459/

PMG (Parliamentary Monitoring Group), 2021. *Research Report on Alexkor Deed of Settlement and Community Structures*, 1 March. https://static.pmg.org.za/210303Brief_on_Alexkor_Deed_of_Settlement_and_Community_Structures.pdf

PMG (Parliamentary Monitoring Group), 2022. *Alexkor & Richtersveld Community Deed of Settlement Progress Report*, 8 June. https://pmg.org.za/committee-meeting/35181/?via=homepage-feature-card
Sida !hub CPA (Communal Property Association), 2007. Annual Report.
Sida !hub CPA (Communal Property Association), 2010. Development Plan 2010. Volume III. Developed by P. Van der Watt & M. Feldman for Sida !hub CPA.

11 Tshikondeni

Mine closure in a deeply rural area

Verna Nel and Dolphin Mabale

11.1 Introduction

The now-closed Tshikondeni coal mine is in the far north of the Limpopo Province, close to the international borders of Zimbabwe and Mozambique and near Kruger National Park. This is a deeply rural, isolated area administered by traditional leaders. The Tshikondeni village had houses and recreational facilities for about 150 mining families. Many residents now survive by subsistence farming, as there are few employment opportunities in the area other than working at tourist lodges and nature reserves. Consequently, rural-to-urban migration occurs. Informal settlements are growing around Musina (formerly known as Messina), the only large urban centre nearby.

The Tshikondeni mine belonged to Exxaro, a black-owned South African mining company with international investments and operations. Mining at Tshikondeni started in 1984, selling coal to ArcelorMittal, a steelmaking company. But by 2011, it was clear that the reserves were becoming exhausted, and the company started a closure process. In the interim, coal was extracted from three new small open-pit mines, whereas previous mining operations had been underground. In September 2014, production ceased and rehabilitation of the environment commenced.

In South Africa, mine closure is complex and difficult to control successfully. The average lifespan of a coal mine may be 30 years, but estimates vary. The eventual closure has direct and indirect environmental and social consequences for landowners and surrounding communities, often protracted and sometimes permanent. Besides land degradation and water pollution, the social harmony of communities is ruptured by unemployment, poverty and crime.

This chapter looks at a variety of views on the mine closure process and its effects on the area's people. We conducted a focus interview with five members of the Traditional Council and individual interviews with 16 community members (5 women and 11 men who had lost their jobs on the mine). We used a thematic analysis to analyse the interview data. As Exxaro declined an invitation to be interviewed, we had to rely on Exxaro's integrated annual reports and research reports. We also consulted other sources, such as newspaper and magazine articles.

Detailed information about Tshikondeni and its transformation, and the area more generally (the Musina Local Municipality and the Vhembe District

DOI: 10.324/9781003403326-11

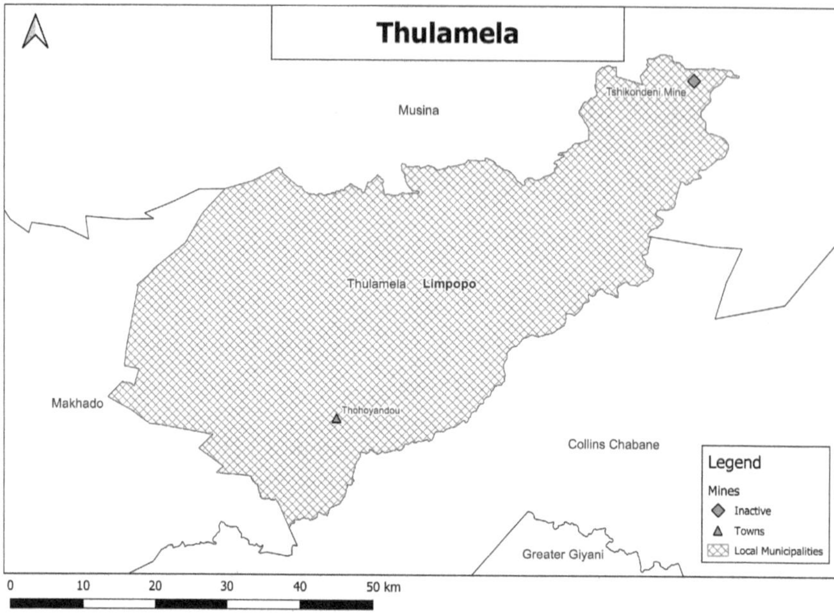

Figure 11.1 Mines and settlements in Thulamela

Municipality, of which it is part), may be found in the following sources: Cornish (2012), Tshitangoni and Francis (2016), Tshikhudu (2018), Machaka (2019) and Odendaal (2019). On the problems of mine closure in South Africa, the difficulties of satisfying everyone's needs and avoiding doing harm, and suggested solutions, see Lane et al. (2015), Tambo and Theobald (2018), Siyongwana and Shabalala (2019), Watson and Olalde (2019), Perkins et al. (2020), Mpanza et al. (2021) and RSA (2022) (see Figure 11.1).

11.2 Exxaro

Exxaro is one of the country's biggest coal producers, and it also has mineral sands and ferrous metal mining operations and is branching into renewable energy (Exxaro, 2020, p. 22). According to its integrated reports, it prides itself on being a socially responsible and environment-friendly company that cares for its workers. For example, in one report, it says its 'approach to sustainability focuses heavily on the physical environment and social communities with which we are involved' (Exxaro, 2014, p. 13). In another, it says it 'acknowledges its moral and legal obligation to ensure responsible and sound environmental management performance, from cradle to grave, for all its operations' (Exxaro, 2011, p. 27).

Interviewed by the Brunswick Group, Mxolisi Mgojo, CEO of Exxaro, spoke about restoring mining land so communities can have a sustainable future:

> [Rehabilitation of the land] has been part of the original mine plan when you start developing your project to open a mine. You should already be thinking about it. Let's take a mine that has a 20-, 30-year life. You don't mine all of it in any one year. So, it's about how you can utilize the areas that would be mined later, so they can actually start economically benefiting the community now. In the past, we used to mine and put stuff into big stockpiles and then return and try to rehabilitate the land later. Now, it's about doing concurrent rehabilitation as you are mining. But with the view of saying: 'We're not going to close this mine; we're going to repurpose this mine into a different economy that has to coexist as we rehabilitate land'.
>
> (Quoted in Mahabane, 2022, pp. 33–34)

The 2012 Integrated Report (Exxaro, 2012) notes that a forum established in 2011 made policy recommendations on the standards and process of mine closure. Exxaro had plans for all the mines that it was closing to be properly closed within five years with adequate financial provision for rehabilitation. Exxaro allocated nearly R400 million for restoring the Tshikondeni and Inyanda mines. Although the more recent Exxaro documents cited here are mostly about generic aspects of mine closure, there are some specific references to the mine closure in Tshikondeni. The emphasis is on environmental rehabilitation, with some reference to social consequences. As part of the closure process, Exxaro says it engages extensively with interested and affected parties to minimise the social impact of the closure. The 2012 Integrated Report refers to studies determining whether the local community can use the existing infrastructure and machinery so the company can leave a positive legacy after closure (Exxaro, 2012, p. 148).

Various laws, such as the Mineral and Petroleum Resources Development Act 28 of 2002 (MPDRA) and the National Environmental Management Act 107 of 1998 (NEMA), regulate the mine closure process and its implementation. The process is subject to a set of standards. It begins with preparing objectives, identifying risks and developing strategies, and ends with implementing the closure plan. The outcome of this process is an approved closure certificate issued by the Chief Inspector of Mines.

In summary, the following are Exxaro's intended principles and actions for the closure process: It says it will develop and carry out a communication plan for affected employees. It will equip employees with skills they can use to find other employment, help host communities acquire business skills and provide opportunities for employees and communities to improve their quality of life. It will devise strategies for engaging with NGOs and communities affected by the closure and ensure the strategies are carried out. It will identify stakeholders' needs and expectations and the potential socio-economic impacts. It will train employees in managing and implementing closure plans and address health and safety issues during the mine closure. It will align the closure with community expectations and honour its

SLP (social and labour plan) commitments. It says closure must be liability-free, and management must be committed to providing resources for implementing the closure plan.

Its 2012 Integrated Report describes the process adopted for Tshikondeni. This entailed forming a community forum as part of the mine closure committee and preparing a database of stakeholders and a communication plan. Research would be done to identify the needs and expectations of stakeholders (including local communities), followed by a socio-economic assessment to determine the impact of the closure. The social and environmental closure plans could be finalised and implemented using this information. The last stage would be obtaining support from stakeholders and approval from the Department of Mineral Resources and Energy (Exxaro, 2012, p. 151).

As part of the closure process, Exxaro engaged with the mine staff on alternative options post-closure, including possible re-skilling. It transferred some staff to other mines. In partnership with the Makuya Traditional Authority and ArcelorMittal, Exxaro included most of the existing infrastructure of the existing village in a trust, the Tshikondeni Legacy Project. This initiative included a photovoltaic minigrid project to ensure electricity was available to the houses during the day and a bio-digester to provide power at night. The CO_2 extracted from the digester could also be sold to bottling plants (Exxaro, 2018a; 2018b). Wood from the land cleared for the microgrid was made available for the community to use or sell as firewood. In 2019, it was reported that there are plans to develop Tshikondeni into an eco-village, with the launch of initiatives such as the Laurelle Zamisa Malaria Research Centre (Odendaal, 2019), which saw the conversion of the old mine hostel.

Of the 298 hectares of disturbed land at Tshikondeni, 139 had been rehabilitated by 2020. The intention is to transfer 90% of this rehabilitated land to emerging farmers by 2026, thus contributing to land redistribution (Exxaro, 2020).

In 2019, Exxaro handed over a community hall, three crèches and 156 houses to the Tshikondeni community (Exxaro, 2018a). In a media article on the launch of Impact Catalyst,[1] mention is made of a post-mining R228 million investment programme by Exxaro (Machaka, 2019). Exxaro also established the Makuya Tshikondeni Development Foundation as a joint venture with ArcelorMittal SA and the Makuya Traditional Authority (Odendaal, 2019).

Kone (2019) corroborates some of the claims in the Exxaro reports. Senior staff and management confirmed that the closure and social labour plans had been followed. Thus, the process met legal requirements. There had been adequate communication with the staff during closure sessions. Literacy[2] and reskilling training was offered, though not all workers took advantage of this opportunity. Employees who wanted to be transferred to other mines were accommodated, and those who did not were given retrenchment packages. Exxaro provided counselling on financial management. However, they acknowledged that the mine could have managed the closure process better, particularly regarding the environmental rehabilitation and the financial counselling for workers. Many suppliers to the mine, such as local small-scale farmers, lost their main client (the mine). The interdependency between them and the mine was thus broken. Because of their dependence on the mining market, small-scale farmers failed to consider other avenues of market access, though in such an isolated area the choice is not wide.

11.3 The traditional leaders' views

Traditional leaders are a powerful intermediary between the local community, government and private business. Tshitangoni and Francis (2016, p. 243) describe this leadership as being 'at the coal face' of local realities. The traditional leadership has played a prominent role in post-apartheid South Africa. The Constitution (RSA, 1996) officially recognises the role and responsibilities of traditional leaders in the lives of many traditional communities that practise customary law. Their legitimacy is supported through the Traditional and Khoi-San Leadership Act 3 of 2019 (TKLA), which establishes councils of traditional leaders at national, provincial and local levels. As most traditional leaders operate at the local level, there can be friction with the municipality and elected officials. The Legal Resources Centre (2021, p. 19) refers to the traditional leaders as 'a fourth tier of government', and it notes that this poses some risks; for example, these leaders may usurp government roles but cannot be held to account as they are not legally elected government officials. An opinion piece in the Bulawayo *Chronicle* notes that traditional leaders near Tshikondeni are concerned about mining companies accessing resources at the expense of the community. Strong traditional leadership is essential to protecting communities against bribery and exploitation (Stevens, 2023).

The five traditional leaders we spoke to took a negative view of how the mine had operated and closed. They believed the community should have benefited more during operations. They thought the closure had caused the local socio-ecological system to deteriorate and forced shops, banks and petrol stations to close. In the following, we give details of six distinct concerns they expressed: the mine's alleged failure to tell them it was closing, insufficient help from the mine, closure of businesses and amenities, social disruption, loss of well-being and damage to the environment.

Despite the mine's claims about informing the community about the closure, the traditional leaders did not think such communication had taken place. One said, for example, that 'there was no announcement of the mine closure by the bosses. No negotiations whatsoever. The closure surprised us all'. It is difficult to pin down the reason for the discrepancy between the mine's assertion and this leader's. It could well be that the message never reached this leader. But this could also be a case of 'cognitive lock-in', making people deaf to the messages about closure.

They thought the mine had done nothing for the surrounding communities while it was still operating. They acknowledged that the company had built two crèches and donated the houses in the mining village after closure. But they claimed that providing the houses had created conflict as the mine had not discussed the allocation criteria with the traditional leaders.

They said the mine closure had deprived the surrounding community of access to economic and social amenities. The village used to have a clinic, a supermarket, a petrol station, a bank, a post office and other amenities that provided services for the mine and the community. These were closed when the mine was closed. People now had to go to Musina or Thohoyandou for their shopping, and the only functional clinic was also some distance away.

They mentioned examples of social disruption following the mine closure. They said alcoholism and prostitution were rife. They said prostitution had become a common survival mechanism for young girls. They said there had been an increase in divorces. They were concerned about increased crime. They said, for example that poaching had become a way of ensuring a livelihood for some former miners.

They said there had been a loss of individual well-being. They said retrenched miners' diets had deteriorated, their physical health had suffered and they were struggling psychologically. They disputed the availability of counselling services.

They expressed concerns about the physical environment. They said some people who had lost their jobs had turned to chopping down trees to sell the wood, which is a crime because it is illegal.

11.4 The community's views

The community respondents' views were only slightly less negative than those of the traditional leaders. The physical and social deterioration we saw in the area is evidence of the rural communities' vulnerability and their dependence on mining. In the following, we give details of the six main concerns they spoke about: notification of closure, help from the mine, loss of mine support and deterioration of lifestyle, loss of jobs and income, financial unpreparedness, social disruption and environmental damage.

Most of these respondents said the mine had given ample notice of the possible mine closure, but this 'did not mean they were prepared for unemployment'. They all said the mine had communicated its closure plans well in advance. Some said the warning had come about two years earlier.

They acknowledged the interventions to support them (counselling, financial advice and the opportunity of reemployment at other mines), although some said the counselling had been in a group format and could have been more helpful. However, most felt that the severance packages had been inadequate. They were less hostile towards the mine than the traditional leaders, but they did say they had suffered since the closure.

They painted a picture of a village that had been prosperous and full of life while the mine was in operation. Even when the mine shifted to contract work, the contractors employed many community members. They mentioned ways in which mining had supported the local community. Some said they had built a house and placed their children in a school of their choice. Since the closure, they said, there has been an outflow of people. They complained of losing access to shops, clinics (despite two new ones being built) and a soccer pitch. The local supermarket, now closed, used to provide credit to mineworkers. Respondents spoke of their lives before the mine closed: 'Life was easy then, but the closure made everything bleak'.

Common remarks we heard were: 'life is tough', 'people are struggling' and 'widespread poverty exists'. Some retrenched miners were concerned about losing the mine-provided medical aid, which would have ensured they got good healthcare, and about not having enough money to ensure their children's education. One reported that a child had had to drop out of tertiary education. Many respondents said they now rely on social grants. They said their severance packages were inadequate for the long term, and they complained about the deduction of tax from the original amount. These responses point to unsuccessful financial counselling on the part of the mine and failure to manage employees' expectations adequately. Some respondents, however, had used the package to generate income: one had opened a brickyard, and another was providing transport services. But we also heard about business failure. The respondents said it had been hard to find a new job. One had used her skill as a chef to bake goods and sell them to a local school. Another said it took four years to find a new job. This is not uncommon where a mine has closed. Path dependency is to blame: mining skills do not easily translate to jobs in other sectors. Having a mining job is often a form of lock-in.

The deterioration of the local economy has had a cascading effect on the area's future and its social cohesion. The respondents spoke of social disruption and crime associated with mine closure. Most of them mentioned an increase in divorce (or couples separating), alcoholism, prostitution and stock theft. They said people were suffering from psychological problems such as stress and loneliness. One respondent said mine closure 'made me sick'. Some recalled local protests and road closures because of the protests, but these had no effect.

They said that infrastructure had been damaged while the mine operated. They thought the mine should have addressed these concerns while it was in operation. It was not until closure that the mine repaired houses that had cracked because of mine blasts, and some respondents complained that the reconstruction of houses was done without consultation. They said the local environment had been damaged by the mining, and this was being made worse by people illegally felling trees to make a living selling firewood. One respondent specifically said that he had started a business selling firewood, and although the authorities require a licence, he operated without one.

We noticed, however, that neither the traditional leaders nor the community respondents mentioned the rehabilitation process or the Tshikondeni Legacy Project and its initiatives. None of them reflected on the attempt to change the mining village into a tourism attraction.

11.5 Conclusion

Cross-tabulating the views of the three sets of stakeholders – the mine, the traditional leaders and the community – we find some areas of agreement and some divergence regarding the successes and failures of the mine closure.

The mining company thought they had followed the letter of the law in closing the mine, and there is no reason to suspect they did not do this. Not only did they focus on land rehabilitation, but they also repurposed the mining village. Yet following the letter of the law and changing the area's economic function cannot entirely counter the hardships that mine closure brings. The views we heard point to a mismatch between what people expect and what the mining sector is legally required to do to ensure sustainable and resilient post-mining environments. Legal provisions do not always prescribe ethical requirements. Best practices need more comprehensiveness, cohesion and transparency and must adequately address end-state responsibilities and liabilities (Perkins et al., 2020). Part of the problem is that best practices and legislation make simple assumptions about post-mining economies and the long-term effects of dependence on a mine. Repurposing the mining village for tourism is a case in point. This was a good idea, but more reflection was needed on the local capabilities and interests. Furthermore, the repurposing was viewed as a mining investment; the mine put the infrastructure in place without considering whether people wanted it or had the capacity to operate it. But this investment differed in one major way from the usual run of mining investments: the amount invested was substantially smaller. Large mining investments change the socio-economic environment and create dependence. Exxaro's small investment after mine closure is unlikely to have that damaging effect.

The closure process is expensive for rights holders, and it is difficult to determine and quantify where liability starts and ends in terms of time frames. This questions the logic and validity of the process and the closure certificates, as post-mining landscapes cannot be rehabilitated to a state where all future risks are eliminated. Closure is a precarious balancing act involving developing the rural economy, satisfying business and investment needs, and protecting the local community from harm. Both present and future needs require careful consideration.

Notes

1 Impact Catalyst is a public-private partnership for socio-economic change established between Anglo American, Exxaro, the Council for Scientific and Industrial Research (CSIR) and World Vision South Africa to facilitate development (CSIR, 2019).
2 Adult basic education and training (ABET).

References

Cornish, L., 2012. Tshikondeni – Almost time to say goodbye: Hot seat. *Inside Mining*, 1 August.

CSIR (Council for Scientific and Industrial Research), 2019. *New Public-Private Partnership for Socioeconomic Change Launched in Limpopo*, 30 October. www.csir.co.za/new-public-private-partnership-socioeconomic-change-launched-limpopo

Exxaro, 2011. *Integrated Report*. https://exxaro-prod.azureedge.net/sitestorage/media/hz4kv0zf/annual-report-2011.pdf

Exxaro, 2012. *Integrated Report*. https://exxaro-prod.azureedge.net/sitestorage/media/h3jlvn2i/integrated-report.pdf

Exxaro, 2014. *Integrated Report*. https://exxaro-prod.azureedge.net/sitestorage/media/5oajd5ep/integrated-report.pdf

Exxaro, 2018a. Post mine closure investment project: Tshikondeni legacy and micro-grid project. *Our Communities*. http://exxaro2018miningindaba.ixd.co.za/our-communities.html

Exxaro, 2018b. *Supplementary Report*. https://exxaro-prod.azureedge.net/sitestorage/media/er5ccyzs/supplementary-report.pdf

Exxaro, 2020. *Environmental, Social and Governance Report.* https://exxaro-prod.azureedge.net/sitestorage/media/25rhai1d/environmental-social-and-governance-report.pdf
Kone, L., 2019. Administration of socio-economic aspects of mine closure. Unpublished doctoral thesis, University of Fort Hare, Ellis.
Lane, A., Guzek, J. & van Antwerpen, W., 2015. Tough choices facing the South African mining industry. *Journal of the Southern African Institute of Mining and Metallurgy*, 115(6), pp. 471–479.
Legal Resources Centre, 2021. *Stop the Bantustans Campaign: Taking the Traditional and Khoi-San Leadership Act 3 of 2019 to Court.* https://lrc.org.za/wp-content/uploads/LRC-TKLA-englishWEB.pdf
Machaka, T., 2019. Impact Catalyst to drive change in Limpopo. *Capricorn FM*, 30 October. www.capricornfm.co.za/impact-catalyst-set-to-drive-change-in-limpopo/
Mahabane, I., 2022. The (un)just transition. *Brunswick Review*, 15 March. www.brunswickgroup.com/mxolisi-mgojo-exxaro-net-zero-i20846/
Mpanza, M., Adam, E., & Moolla, R., 2021. A critical review of the impact of South Africa's mine closure policy and the winding-up process of mining companies. *Journal for Transdisciplinary Research in Southern Africa*, 17(1), pp. a985.
Odendaal, N., 2019. Sustainability initiative transforming former mining village into economic hub. *Miners Weekly*, 19 July. www.miningweekly.com/article/miners-sustainability-initiative-aims-to-transform-former-mining-village-2019-07-05
Perkins, D., Cooper, T., Scholtz, L. & Mulaudzi, K., 2020. Mine closure and rehabilitation in South Africa: Activating coalitions of the willing for a just future. Briefing paper to WWF-South Africa, Cape Town. https://wwfafrica.awsassets.panda.org/downloads/mine_closure_and_rehabilitation_in_south_africa.pdf
RSA (Republic of South Africa), 1996. The constitution of the republic of South Africa (Act No. 108 of 1996). *Government Gazette*.
RSA (Republic of South Africa), 2022. Mine community resettlement guidelines. *Government Gazette*, 22 March.
Siyongwana, P. & Shabalala, A., 2019. The socio-economic impacts of mine closure on local communities: Evidence from Mpumalanga province in South Africa. *GeoJournal*, 84, pp. 367–380.
Stevens, P., 2023. From state capture to Makuya community capture. *Chronicle*, 27 May. www.chronicle.co.zw/from-state-capture-to-makuya-community-capture-ramaphosa-asked-to-intervene/
Tambo, O. & Theobald, S., 2018. Financial provisioning for rehabilitation and mine closure: A study of South African platinum and coal mining companies. *Intellidex, funded by the Centre for Environmental Rights (CER).* https://fulldisclosure.cer.org.za/2018/wp-content/uploads/2018/06/Intellidex-financial-provisioning-for-rehabilitation-and-closure-in-SA-mining.pdf
Tshikhudu, E., 2018. Many jobs to be created as Makuya project starts. *Zoutnet*, 12 November. https://zoutnet.co.za/articles/news/48645/2018-11-12/many-jobs-to-be-created-as-makuya-project-starts
Tshitangoni, M. & Francis, J., 2016. Pillars of resilience of traditional leadership in Vhembe District of Limpopo Province. *Journal of Social Sciences*, 46(3), pp. 241–250.
Watson, I. & Olalde, M., 2019. The state of mine closure in South Africa: What the numbers say. *Journal of the Southern African Institute of Mining and Metallurgy*, 119(7), pp. 639–645.

12 Emalahleni's just transition

From closure to collaboration?

Lucius Botes, Samkelisiwe Khanyile and Zolile Mqotyana

12.1 Introduction and background

South Africa's *Business Day* newspaper says 'SA's climate change ambitions will hurt mining and pose an existential threat' (Erasmus, 2022). In this chapter, we look at some of the problems and how people could get hurt, and we make suggestions for managing the transition to renewable energy with minimal social damage.

South Africa has committed to the Paris Agreement, agreeing to transition away from a coal-driven economy. However, South Africa has some of the highest inequalities in the world. If this is to be a *just* transition, it will have to minimise these inequalities. In 2018, coal exports generated R4.5 billion in revenue and employed about 90,000 people (TIPS, 2021). However, coal is South Africa's biggest contributor to global warming, contributing 80%–95% of CO_2 emissions since the 1950s (Bond & Erion, 2009). Complying with the Paris Agreement will require a shift away from coal. Simon Nicholas, energy finance analyst at the IEEFA (Institute for Energy Economics and Financial Analysis), says:

> New energy technology is replacing coal-fired power faster than most predicted. By 2030, new wind and solar will be cheaper than running existing coal- or gas-fired plants virtually everywhere in the world. This is already the case in India, SA's major coal-export destination.
>
> (Burkhardt, 2019)

The envisaged transition will affect South Africa's coal-fired power stations and its coal exports. Although the country's coal exports have increased since the invasion of Ukraine by Russia, they will probably decline in the long term. Consequently, South Africa is facing serious difficulties. For example, the National Treasury is proposing high carbon taxes (Nel et al., 2023), South Africa's major banks have already announced they will not fund new coal-fired power stations or expansions of old ones (Erasmus, 2022). These changes require a policy response (Burkhardt, 2019).

The geographic location of coal reserves in South Africa exacerbates the problem. Approximately 80% are in Mpumalanga province. There are 111 coal mines in Mpumalanga, of which 50% are active (Nel et al., 2023) and six coal-fired power stations (IASS/IET/CSIR, 2022; Van der Watt & Marais, 2019) (see Figure 12.1).

DOI: 10.324/9781003403326-12

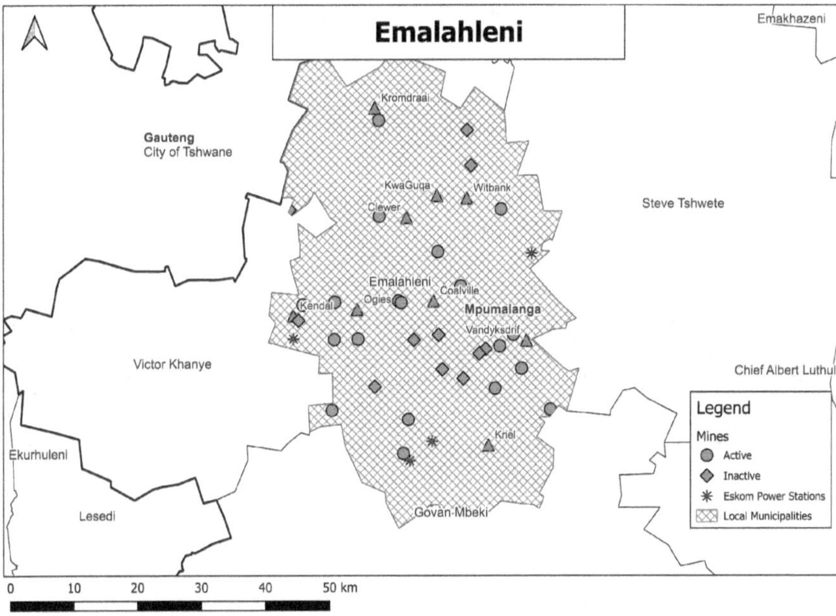

Figure 12.1 Mines and settlements in Emalahleni

Four of these coal-fired power stations, Duvha, Kendal, Kriel and Kusile, are situated in Emalahleni (formerly Witbank), in Mpumalanga. Together, these four power stations generate 15,366 MW (Eskom, 2022). Consequently, Emalahleni's 2021/2022 IDP (integrated development plan) says Emalahleni is the energy heartbeat of South Africa and the economic hub of Mpumalanga. It says coal accounted for 19% of the gross value added of the Mpumalanga province (2021) and increased from R13.6 billion in 1996 to R18.7 billion in 2018, while increasing employment from 20,000 to 35,000 during the same period.

According to Burton et al. (2018), signing the Paris Agreement implies that only the Medupi and Kusile power stations will be in production by 2050. The other coal-fired power stations will close (as they would probably do anyway because most have reached the end of their operational lives), although the country's current energy crisis might mean these power stations will have to stay operational for longer. Moreover, the plan is that coal will contribute only 10% of SA's energy production in 2050, compared to 80% in 2018 (Burton et al., 2018). Therefore, considering the centrality of coal mining and coal-fired power stations in Emalahleni, the energy transition will adversely affect Emalahleni communities. Emalahleni ('place of coal', Zulu) is now facing the challenge of transferring to Bohlabela ('the east', or 'place where the sun is coming from', Northern Sotho). We examined Emalahleni Local Municipality's IDPs for the years 2018/2019, 2019/2020, 2020/2021 and 2021/2022 and a range of government documents associated with the transition.

12.2 Economic, political and social consequences of mine closure

Besides planning to repair damage to infrastructure and the environment, policymakers must decide how to deal with economic problems, such as labour market changes and increased unemployment; political problems, such as governance and service delivery issues and social problems, such as deterioration in health, education and household welfare. A just transition will have to go beyond the narrow issue of the plight of retrenched workers and the physical issue of the environment to include a broader focus on communities and institutions (Marais, Burger et al., 2022).

Vivoda et al. (2019) identify 14 social impacts of mine closure: the economy, business, employment, security, education and training infrastructure, amenity, livelihoods, land, housing, health, environment, demography, participation and inclusion. Marais, Ndaguba et al. (2022, p. 37) add another: social disruption. They expect four major impacts of mine closure to be relevant to Emalahleni. Extensive land damage results from the extraction of mineral ores; water sources are polluted by acid mine drainage and disasters are caused by broken tailings dams. Markets are damaged by unemployment, the collapse of the housing market and the closure of businesses in the mining value chain. Services formerly provided or subsidised by the mines, such as schools, clinics and hospitals, are lost, and government institutions are expected to take over. And social disruption occurs, bringing an increase in crime.

12.3 Policy responses

The Presidential Climate Commission (PCC) has released for discussion several policy documents on a just transition to sustainability, most recently 'Laying the foundation for a just transition framework for South Africa' (PCC, 2022a) and the 'Just transition framework for South Africa' (PCC, 2022b). The February 2022 report benchmarks existing work in the knowledge area of just transitions in South Africa, and the July 2022 report is the first draft of a just transition framework for the country that 'will be a strategic plan for achieving a just and equitable transition to net-zero CO_2 emissions in South Africa by 2050, setting out the actions that social partners will take over the short, medium, and long term' (PCC, 2022a, p. 6). South Africa has secured US$8.5 billion to support its just transition, including repowering and repurposing decommissioned coal-fired power stations (Stats SA, 2023). However, the just transition framework of 2022 acknowledges that South Africa is one of the most unequal countries in the world, complicating the transition from coal. The high levels of social stratification pose many risks, especially considering that South Africa is a young democracy with weak institutions (Von Holdt, 2013).

Five key policy options have been suggested that could help achieve a just transition in Emalahleni and Mpumalanga as a whole:

- *Investing more in clean energy*: As the province already has a wealth of experience in the business of energy creation, by investing more in renewables, it could become the green energy hub of South Africa.

- *Diversifying the economy*: The shift to renewable energy would create some jobs for unskilled and semi-skilled labour, but a net job loss is expected in the province by 2030. To be just, the transition would therefore have to be accompanied by economic diversification and investment in other sectors such as tourism and agriculture (agro-processing).
- *Improving the position of women*: The transition from fossil fuels to clean energy could be an opportunity to improve representation and conditions for women in the energy sector. In the past, they comprised only a small proportion of the coal mining sector: only 29% of the workforce of Eskom (the state electricity utility) and 10% of the coal sector (IASS/IET/CSIR, 2022).
- *Making a regional effort*: As mining affects an entire region, planning for mine closure would need to be a collaborative regional development effort (DMRE, 2021).
- *Reducing dependence on mining*: Importantly, policymakers would have to acknowledge mining dependencies and try to reduce them.

Thus, a drive towards a just transition would entail identifying and preparing institutions to drive the transfer to renewable energy and away from coal-driven energy. A well-functioning local municipality will be essential; the currently dysfunctional Emalahleni municipality would not be able to plan and manage a post-mining region successfully. A thriving business chamber will be needed to help people setting up new businesses. Good intergovernmental relations will be needed between the various departments and spheres of government. Kanyane (2016) notes that good intergovernmental relations ensure cooperative governance and accountability in transitioning towards a post-mining dispensation. Public-private partnerships will be needed to diversify the economic base. Civil society, civil society organisations and community-based organisations will need to be strengthened. Early attempts should be made to learn from any similar coal-mining towns that manage to transition away from coal successfully or from any local economy dominated by a single sector that has managed to diversify its economy.

Chapter 3 noted that mine closure policies and legislation require strengthening, and departments must develop the capacity to implement the requirements. For example, although the Minerals, Petroleum and Resources Development Act (MPRDA) (28 of 2002) requires mines to develop social and labour plans (SLPs), the issue of mine closure needs more attention. An SLP is supposed to dovetail with a municipality's IDP, but often the two do not align as they should. This weakens the relationship between the mines and the municipality (Van der Watt & Marais, 2021). It will be essential for Emalahleni to improve this relationship to ensure cooperative planning for a just transition.

The CoBenefits report (IASS/IET/CSIR, 2022) suggests that South Africa could become a renewable energy 'superpower' – it ranks sixth worldwide in renewable energy production potential per square kilometre, largely because of its favourable climate. Renewable energy production is a major source of jobs in South Africa. The report notes that the government should be doing as much as possible in the policy space to allow this industry to flourish while at the same time managing the transition from fossil fuels through targeted programmes, such as those aimed at reskilling people currently employed in the coal mining industry. However, this is not happening in Mpumalanga, according to Nel et al. (2023).

Complicating the transition, South Africa's Integrated Resource Plan (IRP) (DMRE, 2019) still has what Joubert (2019) calls a 'religious commitment' to drive the country's energy with coal. Emalahleni's IDP, like the IRP, hardly refers to the back-end plans for transitioning from coal to renewable energy and the eventual decline and closure of the country's coal mines, despite the country's suitability for renewable energy generation.

We believe it is likely that transitioning to renewable energy will not be just a linear case of job loss but that there will be some scope in the renewable energy sector for low-skilled and semi-skilled workers who have been retrenched from the coal mining sector. Some reports, such as the CoBenefits report (IASS/IET/CSIR, 2022), focus on retaining the skills of existing workers, as the bulk of the jobs will be for construction, installation, generation, maintenance and wholesale in the renewable energy sector. According to this report, 68,000 to 72,000 jobs could be created if an ambitious decarbonisation scenario in Mpumalanga is realised.

The CoBenefits report points to some significant benefits of transitioning to renewable energy generation. By repurposing some of the property, plant and equipment used in the coal mining industry, Mpumalanga could compensate for some of the socioeconomic damage caused by mine closure, including a large share of job losses in the declining sector. By investing in the transition, Mpumalanga could lay the foundations for becoming the country's new clean energy hub, and the province's gross output value could be substantially increased. However, this will require significant commitment and financial outlays from various stakeholders (IASS/IET/CSIR, 2022). For example, the transfer of skills from coal mining to renewable energy generation must be a process that is included in a mine's life cycle. Skills development programmes in preparation for a just transition should not be an afterthought when mining operations have ceased.

A well-functioning local institutional context is vital to making plans that will be viable for a post-mining economy. Good social capital, in the form of cooperation, trust and collective decision-making capacity, is needed for local development. The institutions in Emalahleni and Mpumalanga need the capacity to manage mine closures and the closing of coal-fired power stations. Marais, Burger et al. (2022) describe the ways in which policy fails to tackle the impacts of mine closure: insufficient global and country-level knowledge, better management of environmental than social impacts, and an excessively front-end approach to a mine's life-cycle. Mining legislation is part of environmental, water and pollution legislation, and no legislation or regulation exists that focuses exclusively on mine closure (Crous et al., 2021).

Emalahleni Municipality's 2018/2019 IDP has no strategic objectives for preparing for mine downscaling. It says nothing about the shift from coal energy to renewable energy or the urgency of diversifying the economy, and it mentions no plans for dealing with mine closure. It makes only two references to mining issues. The first is about the limited opportunities to involve small artisanal mining operators in current and new mine ventures because heavy machinery is needed for coal mining, and the second is about the shortage of land for smallholder farmers due to mining operations. This IDP seems to be an inward-looking, state-driven document where service delivery issues take precedence. However, the 2021/2022 IDP firmly acknowledges, in the mayor's preface and the municipal manager's message, the challenges Emalahleni will face in moving away from coal to green energy. But again, it mentions no concrete strategic goals or objectives to achieve this transition. Reskilling, reusing infrastructure and facilities and seeking new employment options are not yet happening in Emalahleni (Nel et al., 2023).

There are three prerequisites for a just transition in Emalahleni: capacity, efficiency and integrity. Capacity must be increased. The involvement of politicians, bureaucrats, businesspeople and community leaders is crucial. Trust networks must be created between these stakeholders to render social capital a strategic resource to enhance local competitiveness. Provincial, district, and local spheres must become more efficient. This will not be achieved easily. For example, the Auditor-General (2022) revealed that the Emalahleni Municipality owed Eskom R5 billion in June 2021 and is one of the top three defaulting municipalities in South Africa. Local development depends on the capacity to learn and develop specialised knowledge areas. For this to happen, the patrimonial state in Emalahleni and Mpumalanga must be exposed and broken. Local government officials are largely influenced in their actions and decision-making by personal relationships with those with political power and ties to businesspeople based on the appropriation of public resources (Stevens, 2019; Trigilia, 2001).

12.4 The likelihood of a just transition

To be just, the transition to renewable energies should not come at the expense of people. The Draft National Mine Closure Strategy (DMRE, 2021) makes provisions for post-closure stewardship, socioeconomic sustainability and the ongoing monitoring of closure plans. The strategy seeks to mitigate the risks of closure in the earliest phases of the mining cycle. Clear objectives are to be set out for each phase of a mine's development. Planning must take into account the impacts and prospects for the economy, the community and the environment. Scenario planning must be done to align a forecasted future with a preferred future, anticipating the pros and cons of mine's decline.

Although the disadvantages of closing the coal mines in Emalahleni far outnumber the advantages, there may be benefits in the long term, such as a reduction in air pollution and therefore health benefits (Denoon-Stevens & Du Toit, 2022); a reduction in water usage, as Eskom is the single largest consumer of raw water in South Africa (Bond & Erion, 2009); an improvement in the area's ecology (though there will be problems with changes in land use and the reintegration of previously mined land) and a reduction in the impact of extractions and excavations on the land and local communities.

However, according to Marais, Burger et al. (2022), these benefits are only realisable if a broader view of the impact of a declining mining sector is taken. Role players must consider the short- and long-term consequences. They must look at the broader community implications and move away from a narrow focus on miners and environmental impacts. Managing mine closure and transitioning to renewable energy will require competent governance and institutions. The Draft National Mine Closure Strategy seeks to develop regional closure strategies for clusters of mining operations. This will require holistic post-mining management. There is also a need to learn from the 'shrinking cities' literature, which focuses on the possible benefits of decline (Marais & De Lange, 2021). If there are to be benefits, the transition must prioritise alternatives for workers who will lose their jobs in coal mining and coal-fired power stations.

The Draft National Mining Closure Strategy wants to restore post-mining land to pre-mining conditions. It suggests that all role-players should plan mine closure and include it in all phases of the mining cycle. The strategy says mining operation applications must comprise business plans and be accompanied by SLPs when mining operations commence. This would encourage the inclusion of communities and multiple stakeholders in the mine life cycle, which could fast-track the diversification of Emalahleni's economy. However, there are few examples of successful economic diversification in post-mining economies (Marais, Burger et al., 2022). There is a need for economic benchmarking, which is critical to understanding the full economic impact of mining on mining-dependent communities (DMRE, 2021).

Many studies on mine closure focus on the mining company's behaviour and say less about the behaviour of the state, the community, the business sector and other stakeholders. In planning for a post-mining economy, urban inclusivity (Magidimisha-Chipungu & Chipungu, 2021) and a participatory planning approach (with all its benefits: an increased sense of ownership, drive and envisaged future) will be essential. Stakeholders such as mining companies, Emalahleni businesses, community groups, the provincial and local governments, environmentalists and trade unions need to come together to drive a coordinated and dedicated front-end approach to a post-coal economy in Emalahleni. There is a need for a broad-based forum with dedicated champions driving it. These Emalahleni stakeholders need to understand the importance of collaboration. Its advantages will benefit not just individuals but also everyone participating in the forum.

12.4.1 Dependencies

South Africa's coal mining industry has generated wealth for mining companies and is often cited as an economic multiplier, promoting industrialisation. It has created interdependencies between the mining sector, the state, the market, the private sector and society. According to the PCC (2022a), coal mines employed around 80,000 across the country in 2021, declining from 140,000 at the sector's peak in the 1980s. There are currently over 100,000 workers in Mpumalanga in the coal and related industries (Nel et al., 2023). In 2020, the mining sector accounted for 22% of employment and 48% of the gross value added (Global Insight, 2021). In Emalahleni, coal mining contributed 45.8% to GDP in 2021 (see Table 12.1). Although coal mining operations have been mechanised since the 1980s, resulting

Table 12.1 Contribution of economic sectors to Emalahleni's GDP, 2011 and 2021

Sector	*2011*	*2021*
Agriculture	0.6%	0.7%
Mining	42.7%	45.8%
Manufacturing	9.8%	9.4%
Electricity	8.7%	7.5%
Trade	10.5%	9.6%
Construction	2.3%	1.7%
Transport	4.6%	4.0%
Finance	9.0%	9.4%
Community services	12.0%	11.8%
Total industries	100.0%	100.0%

Source: Global Insight (2021)

in a decline in employment, the economy of Emalahleni is still dominated by coal mining and related activity (Campbell et al. 2017). Coal mining's contribution increased by 3% between 2011 and 2021 (see Table 12.1).

Emalahleni has a TRES Index score of 57.9, while the national TRES figure is 40.3 (Global Insight, 2021). TRES measures the level of diversification or concentration of a region's economy. A score of 0 represents a diversified economy (meaning all economic sectors contribute equally to the GVA of a region), while a score of 1 reflects an economy dependent on one sector only. This suggests that the economy of Emalahleni is not very diverse and is heavily skewed towards one sector, the mining sector. The more diverse an economy is, the more likely it is to create employment opportunities across all skill levels and maintain a balance between labour-intensive and capital-intensive industries.

Various dependencies can be identified in Emalahleni: interdependency, goal dependency and path dependency. There are several interdependencies between the mining company and the local institutions. We can see goal dependency in the government's drive to enable private homeownership for mine employees, which may render these households even more vulnerable during mine decline if the local economy declines and their houses lose value. The mines attracted an influx of people to Emalahleni, resulting in the growth of informal settlements. One in four houses in Emalahleni is informal, outstripping the national percentage of 14% (Marais, Burger et al., 2022).

The mines in Emalahleni have invested heavily in the urban infrastructure. As the mines scale down, the mine-provided roads, water supply, sport and recreation facilities, educational facilities, public spaces, and so on will become the municipality's responsibility. Some mines currently provide water to the municipality, further rendering Emalahleni and its communities vulnerable if these mines close. These multiple path dependencies often lead to an incapacitated local municipality not being able to care for and capitalise on using these mining-provided assets to the greater benefit of the community and helping Emalahleni to a more socioeconomically and environmentally viable post-mining situation.

12.4.2 Social disruption

The decline of the coal mining industry is likely to bring social disruption to Emalahleni, in the form of unemployment, increasing inequality and endemic poverty. Job losses will be heavy inside and outside the coal mining industry. The Gini coefficient for Emalahleni remained constant at 0.6 from 2011 to 2021 (Global Insight, 2021).

Many non-mining households (with no members employed on the mines) benefited directly and indirectly from the wealth created by the mining operations. If the mining operations decline, Emalahleni's large rental market may collapse, and informal settlements may continue to grow (Marais, Burger et al., 2022). Mine closures will also seriously reduce municipal income because people rely on income from the mines to pay rent and taxes. Emalahleni's finances are already in a dismal state, and a mining decline will exacerbate the current crisis.

Mpumalanga's future is not clear. The province's unemployment rate was 34.7% in 2018, compared to 29% for the rest of South Africa (Cronje, 2019). Without a purposeful and strategic plan for dealing with the risk of mine decline and the gradual phasing out of coal, there will be substantial economic losses and enormous socio-economic disruptions. However, repurposing coal-fired power stations through the development of renewable energies (e.g. at the Komati Power Station) can help mitigate the social impacts of mine closure.

12.5 Conclusion

This chapter discussed the difficulties of achieving a just transition from coal-driven energy to renewable energy in Emalahleni, a dominant coal region of South Africa. It looked at the potential socioeconomic impacts of the downscaling of the coal mining sector and the policies that have been suggested for making the transition. This chapter provides an overview of the importance of collaboration between stakeholders.

The transition from fossil fuels such as coal towards wind, solar, hydro and other renewable energy sources is imminent. The challenge remains: how to make the Paris Agreement a success for the people of Emalahleni. The transition should not be aimed merely at achieving goals for mitigating climate change. It should not be climate versus community, but climate *and* community. Moving from coal to green must be meaningful. We cannot plan for a post-mining future without good governance and committed coordination. The chapter has made some suggestions as to how the stakeholders at Emalahleni might plan for life after coal. A just transition from coal to renewables is a complex and difficult endeavour, though essential and challenging. It will have to take into account the many institutional dependencies that a mine-dominating economy brings and also acknowledge and attempt to mitigate the many social disruptions a predominantly mining economy will leave as a legacy once the transition to a post-mining economy happens. This will remain a challenging task and wicked problem. It reminds us of the words of George Monbiot, *Guardian* columnist and environmental and political activist: 'Despair is the state we fall into when our imagination fails. When we have no stories that describe the present and ground the future, hope evaporates. Political failure is, in essence, a failure of imagination' (Monbiot, 2017). We hope our policymakers will not fail to imagine a post-mining future for Emalahleni.

References

Auditor-General, 2022. Emalahleni: Audited financial statements. Pretoria: Auditor-General.

Bond, P. & Erion, G., 2009. South Africa carbon trading: A counterproductive climate change strategy. In McDonald, D. (ed.), *Electric Capitalism: Recolonising Africa on the Power Grid*. London: Earthscan, pp. 338–358.

Burkhardt, P., 2019. *South African Coal Exports Seen Waning as Renewables Gather Pace*. New York: Bloomberg Markets.

Burton, J., Caetano, T. & McCall, B., 2018. Coal transitions in South Africa: Understanding the implications of a CO_2-compatible coal phase-out plan for South Africa. *Energy Research Centre*. Cape Town: University of Cape Town.

Campbell, M., Nel, V. & Mpambukeli, T., 2017. A thriving coal mining city in crisis? The governance and spatial planning challenges at Witbank, South Africa. *Land Use Policy*, 62, pp. 223-231.

Cronje, F., 2019. *The Whole Nine Yards: Profiling South Africa's Provinces*. Johannesburg: Centre for Risk Analysis.

Crous, C., Owen, J.R., Marais, L., Khanyile, S. & Kemp, D., 2021. Public disclosure of mine closures by listed South African mining companies. *Corporate Social Responsibility and Environmental Management*, 28(3), pp. 1032–1042. https://doi.org/10.1002/csr.2103

Denoon-Stevens, S.P. & Du Toit, K., 2022. The health impacts of coal mining and coal-based energy. In Marais, L., Burger, P., Campbell, M., Denoon-Stevens, S.P. & Van Rooyen D. (eds.), *Coal and Energy in South Africa: Considering a Just Transition*. Edinburgh: Edinburgh University Press, pp. 102–119.

DMRE (Department of Mineral Resources and Energy), 2019. *Integrated Resource Plan*. Pretoria: DMRE.

DMRE (Department of Mineral Resources and Energy), 2021. *Draft National Mine Closure Strategy*. Report no. 446. Pretoria: Department of Mineral Resources and Energy.

Erasmus, D., 2022. SA's climate ambitions will cost mining sector, says Minerals Council. *Business Day*, 10 May, p. 11.

Eskom, 2022. Coal fire stations. https://www.eskom.co.za/eskom-divisions/gx/coal-fired-power-stations/

Global Insight, 2021. *Rex Explorer*. Pretoria: Global Insight.

IASS/IET/CSIR, 2022. From coal to renewables in Mpumalanga: Employment effects, opportunities for local value creation, skills requirements, and gender-inclusiveness. Assessing the co-benefits of decarbonising South Africa's power sector. *COBENEFITS Executive Report*. Potsdam/Pretoria. www.cobenefits.info

Joubert, L., 2019. "Religious commitment" to coal could scuttle SA economy. *Daily Maverick*, 17 September.

Kanyane, M.H., 2016. Interplay of intergovernmental relations conundrum. In Plaatjies, D., Chitiga-Mabugu, M., Hongoro, C., Meyiwa, T., Nkondo M. & Francis Nyamnjoh, F. (eds.), *State of the Nation – South Africa 2016: Who Is In Charge? Mandates, Accountability and Constellations in the South African State*. Pretoria: HSRC Press, pp. 91–106.

Magidimisha-Chipungu, H.H. & Chipungu, L., 2021. *Urban Inclusivity in Southern Africa*. Cham: Springer.

Marais, L., Burger, P., Campbell, M., Denoon-Stevens, S.P. & Van Rooyen, D., 2022. *Coal and Energy in South Africa: Considering a Just Transition*. Edinburgh: Edinburgh University Press.

Marais, L., & De Lange, A., 2021. Anticipating and planning for mine closure in South Africa. *Futures*, 125, p. 102669

Marais, L., Ndaguba, E., Mmbadi, E., Cloete, J. & Lenka, M., 2022. Mine closure, social disruption and crime in South Africa. *The Geographical Journal*, 188(3), pp. 383–400.

Monbiot, G., 2017. *Out of the Wreckage: A New Politics for an Age of Crisis*. London: Verso.

Nel, E., Marais, L. & Mqotyana, Z., 2023. The regional implications of just transition in the world's most coal-dependent economy: The case of Mpumalanga, South Africa. *Frontiers in Sustainable Cities*, 4, p. 1059312.

PCC (Presidential Climate Commission), 2022a. Laying the foundation for a just transition framework for South Africa. *Pretoria: Presidential Climate Commission*. February. www.climatecommission.org.za/publications/laying-the-foundation-for-a-just-transition-framework-for-south-africa

PCC (Presidential Climate Commission), 2022b. *Just Transition Framework for South Africa*. July. Pretoria: Presidential Climate Commission. www.climatecommission.org.za/just-transition-framework

Stats SA (Statistics South Africa), 2023. *Sustainable Development Goal (SDG) 7: Ensure Access to Affordable, Reliable, Sustainable and Modern Energy for All*. Pretoria: Stats SA.

Stevens, C., 2019. *Provincial Governments are Not Intervening When They are Supposed To: The Case of Emalahleni Local Municipality*. Cape Town: Dullah Omar Institute.

TIPS, 2021. *Beyond Coal. Opportunities to Diversify the Economies of Emalahleni and Steve Tshwete Local Municipalities*. Pretoria: TIPS.

Trigilia, C., 2001. Social capital and local development. *European Journal of Social Theory*, 4(4), pp. 427–442.

Van der Watt, P. & Marais, L., 2019. Normalising mining company towns in Emalahleni, South Africa. *The Extractive Industries and Society*, 6(4), pp. 1205–1214.

Van der Watt, P. & Marais, L., 2021. Implementing social and labour plans in South Africa: Reflections on collaborative planning in the mining industry. *Resources Policy*, 71, p. 101984.

Vivoda, V., Kemp, D. & Owen, J., 2019. Regulating the social aspects of mine in three Australian states. *Journal for Energy and Natural Resources Law*, 37(4), pp. 405–424.

Von Holdt, K., 2013. South Africa: The transition to violent democracy. *Review of African Political Economy*, 40(138), pp. 589–604.

13 Rustenburg

The struggle to go beyond mining

Sethulego Matebesi and John Ntema

13.1 Introduction

Rustenburg is the world's leading producer of platinum, accounting for over 80% of global production. For many decades, the demand for platinum increased, largely for catalytic converters in petrol- and diesel-driven vehicles, but the 2008–2009 global financial crisis slowed it down. In 2012, the strikes at Rustenburg's platinum mines that led to the Marikana Massacre caused instability and again affected production. There are further signs of decline: mine employment has fallen drastically, and mines have changed ownership. In 2020, in an effort to mitigate the economic impact of mine downscaling, Rustenburg devised a long-term plan, 'Rustenburg beyond mining', which calls for repositioning this local municipality as a regional hub for agriculture, sports, health, education and renewable energy (RLM, 2021). According to a former mayor, the municipality 'started getting worried as far back as 2011', but nothing tangible has yet resulted (Tau, 2020).

The growing body of work on the consequences of mining downscaling in Rustenburg points to the resource curse, the disruption of the social fabric, widespread unemployment and high levels of informality (see e.g. Marais, Ntema et al., 2021; Oranje et al., 2021). Van der Watt and Marais (2021) argue that collaborative planning between mining companies, the government and local communities is critical for mitigating the social damage done by downscaling. Collaborative planning is difficult but possible. The Royal Bafokeng Nation, traditionally governed Setswana-speaking community of 150,000 people living in 29 villages around the mines, argues that with 'long-term and evidence-based planning' (Royal Bafokeng Nation Operations Room, 2023), local actors and institutions can cushion the consequences of mine downscaling. But mine closure is even affecting this attempt by the Royal Bafokeng Nation. Collaboration during growth is difficult, but it even becomes harder when mine closure sets in.

In this chapter, we investigate the consequences of mine downscaling in Rustenburg (see Figure 13.1). We interviewed a wide range of stakeholders: municipal officials from the directorates of Policy and Research, Local Economic Development, Enterprise Development and Rural Development; the provincial organiser of Mining Affected Communities United in Action (MACUA); a representative of the Royal Bafokeng Nation Development Trust (RBNDT); and business owners and community leaders.

DOI: 10.324/9781003403326-13

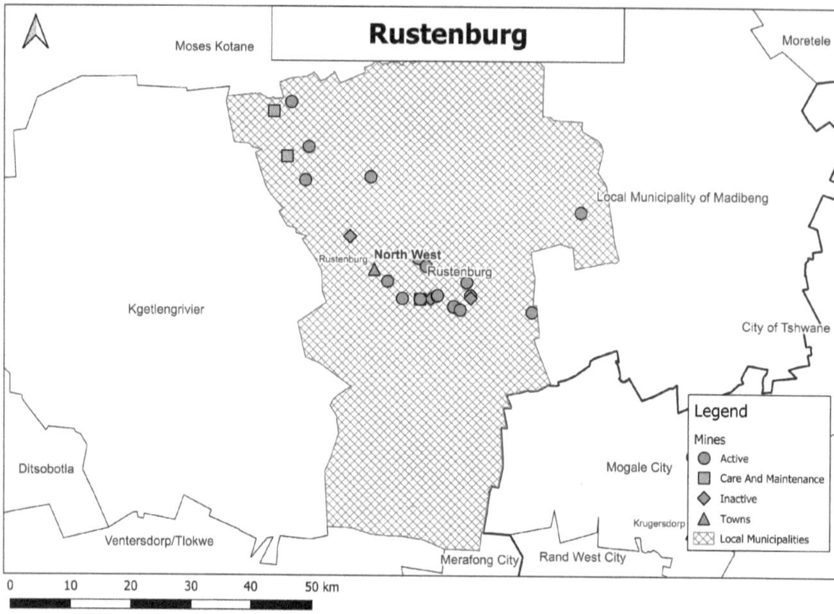

Figure 13.1 Mines and settlements in Rustenburg

We also examined municipal documents. We argue that Rustenburg Local Municipality underestimates the realities of mine closure. Mine closure will make collaboration difficult and will result in new dependencies and social disruption.

13.2 The scale of mine decline and closure

Rustenburg's economy grew by nearly 4% annually between 1996 and 2001 but by only 0.7% between 1996 and 2018 (Marais, Campbell et al., 2021). Further decline is evident in the ownership changes and the reduction in the mine-related labour force. Mine employment dropped from 96,000 in 2012 to 75,000 in 2019 (Marais, Campbell et al., 2021) and to 68,200 in 2020 (RLM, 2022). During the COVID-19 lockdown of 2020–2021, these figures stabilised, and there was even a small increase in employment as mining prospered during the pandemic (largely because the mining sector was not subjected to the most stringent restrictions). The municipality's integrated development plans (IDPs) acknowledge that the shrinking mining sector has affected Rustenburg negatively, using terms like 'declining mining economy', 'undiversified economy', 'growing unemployment' and 'decrease in revenue collection' and mentioning an increase in the number of people in the indigent register (RLM, 2021, 2022). Senior municipal officials who we interviewed painted a general picture of rapidly increasing decline and social disruption, with one saying:

> We now suddenly see that unlicensed shebeens or tuckshops are being run from most houses, while every street in the city has become a trading spot for informal street vendors . . . something we never experienced at the current rate before.

Rustenburg's job-shedding and social instability should be understood within a broader context. Three factors are largely responsible. First, after the global financial crisis, there was an oversupply of platinum, leading to a decline in demand. Platinum production declined by 12%, or about 550,000 ounces, between 2008 and 2012, and there was an 8% drop in global output (Marais, Campbell et al., 2021). Although prices have recovered, volatility remains a problem. Second, the platinum mines in Rustenburg are competing with new open-cast platinum mines to the north, whose production costs are lower. And third, the pressure to switch to electric vehicles will lead to a decline in demand for platinum.

The social aspects of mine closure tend to be neglected by governments and mining companies (Vivoda et al., 2019). South Africa is no exception. For various reasons, the government struggles to close mines: insufficient capability to enforce stringent regulatory regimes for mine closure; unclear or incomplete regulatory regimes for mine closure (Crous et al., 2020); and the failure of South African national policy and local strategies to anticipate the long-term reality of mine closure (Marais & De Lange, 2021).

Like other mining towns in South Africa, Rustenburg's long mining history, with the influx of people and the authorities' inability to manage either growth or decline, is one of fractured families, a damaged social fabric, social disruption and health problems (Lagercrantz & Khabbaz, 2019).

Rustenburg differs, however, in being home to the Royal Bafokeng Nation, which has generated substantial economic benefits through mining. The RBNDT, administered by the Royal Bafokeng Traditional Authority, is South Africa's largest community-based investment company (Mbenga & Manson, 2010; Cook, 2013; and see Comaroff & Comaroff, 2009, for further detail on this re-invention of a culture as a company).

The Royal Bafokeng Nation has shown how mining communities worldwide could ensure local and long-term benefits within the broader context of mine downscaling or closure. This contrasts with Ravensthorpe, Australia, where the mining company completely distanced itself from the community (Mayes, 2014). The Royal Bafokeng Nation is noted for its investment in education. In 2017, for example it established the Royal Bafokeng Institute, which offers youth technical, vocational and educational training programmes (Matebesi, 2020).

Internationally, there is an emphasis on the central role of municipalities in trilateral problem-solving involving mining companies, government and communities (World Bank & IFC, 2002). Municipalities have assisted mine closures in Canada, Australia, Papua New Guinea, Romania, China and Finland (De Lange, 2019). Mitigation strategies have included designing and implementing economic diversification measures. To follow suit, Rustenburg Local Municipality has begun to formulate its Vision 2040, which aims to make Rustenburg 'a world-class city where all communities enjoy a high quality of life' (RLM, 2021). Related to this vision are the city's long-term plan, 'Rustenburg beyond mining', which aims to mitigate the devastating economic impact of further shrinking in the mining sector, and its '2025 diversification vision' (RLM, 2021). But despite these plans, Rustenburg continues to deteriorate, as our interviewees explained and as was evident from the municipal documents we examined. If it fails to put these plans into action, its overreliance on the mining economy will continue.

13.3 Consequences of closure

As a result of the drop in demand and thus in output, Rustenburg Local Municipality is now generating less income and needs help to provide basic services (RLM, 2021). In 2016, revenue collection (in real terms) had declined by 18% since 2012 (RLM, 2021) and in 2020, it had declined by 27% since 2012 (RLM, 2021). This is consistent with a decline in employed people from 245,000 in 2016 to 216,000 in 2020 (RLM, 2022). These trends are largely responsible for the current R5 billion owed to the municipality in rates and taxes, according to a municipal spokesperson speaking to SABC News. He attributed the debt to an 'an increase of people who no longer enjoy formal employment due to the retrenchments in the mining sector' and said 'this affects our ability as an institution' (Moerane, 2020). The increase in mine retrenchments has contributed to an increase in informal settlements in and around Rustenburg. One respondent saw evidence of this in the increase in municipal provision of water via water tankers. The rapid increase in informal settlements often overwhelms the normal water services.

The Royal Bafokeng Nation has also felt the effects of mining closures and the growing unemployment rate. The decline in platinum shares reduced its administration's budget from R1,192 billion in 2009 (Molotlegi, 2009) to R597 million in 2019 (Mosiane, 2021). A municipal respondent said:

> Even developments and amenities that used to be sustained and maintained as assets for the Royal Bafokeng Nation through investment and support or royalties by mines are no longer sustainable due to mine closures and are gradually becoming a liability or burden to us. . . . We are now receiving requests from the Royal Bafokeng Nation to include their community in our local economic development and other strategic planning sessions. . . . All this comes at an extra financial and administrative cost to us as a municipality.

Mining used to be a benefit, enabling substantial improvements in areas managed by Bafokeng. With the decline, it has become a liability for Bafokeng and for Rustenburg.

The growing unemployment rate in Rustenburg has sparked community protests against the municipality and the mines. A live broadcast on SABC News reported a community march to the doorsteps of Anglo-American offices on 26 September 2022 (Kgajane, 2022), and similar community marches have been organised against the municipality. One municipal respondent said that

> instead of marching to the offices of these mining companies, we now see a new phenomenon where some former local miners organise themselves in 'construction mafia' and begin to lead community protests to offices of the municipality, demanding jobs in mines or municipal construction tenders.

The reference to the 'mafia' is further evidence of social disruption closely linked to mine closure.

13.4 Civil society views

Rustenburg has one of the most successful models of local empowerment – but of a single social group, the Bafokeng. This could be to the detriment of others. A respondent from civil society raised concerns about Bafokeng benefiting substantially more than the rest of the population. The fact that the Royal Bafokeng Nation has benefited extensively from mining does not mean that 'the entire Rustenburg community will be able to deal adequately with the impacts of mine closure'. A further concern was that mining companies have concluded royalty agreements with traditional authorities or select community members at the expense of broader mining-affected communities.

Concerns were also expressed that the regulatory environment and practices create more conflict. For example, a community leader said the current financial benefits 'will not be able to offset the existential impacts of mining concerning environmental damage and the well-being of workers and those living in the proximity of mines'. Another community leader felt that a significant governance reality in Rustenburg is that 'traditional leaders perpetuate master-servant relationships'.

However, the representative from the RBNDT disagreed. He said that those making these comments 'conveniently ignore' the realities in which the Royal Bafokeng Nation operates. He said the problem is the short-term orientation that administrators adopt to appease community expectations. In contrast, the RBNDT takes a long-term approach. He said that, however well justified, investment in job creation projects (outside the formal economy) further depresses the financial viability of community trusts and thus makes communities even less able to respond to the shocks of mine closure. He said the RBNDT had planned for life beyond mining using the trust's diversified investment portfolio and investment in education. In his words, the RBNDT believes that 'in straitened economic circumstances, the contribution to the family economy by those who are educated can be considerable'. He said education's longer-term social and economic benefits (in the form of not only income but also resilience) would mitigate the effects of mine downscaling or closure. The Royal Bafokeng Nation has taken a long-term and intergenerational approach to mining and possible mine closure. There is no guarantee that it will be safeguarded against closure, but it has a much better chance than those who take the short-term approach.

13.5 The municipality's views

Senior municipal officials were aware that Rustenburg was undergoing mine downscaling. One of them said, 'we see it, we read about it . . . despite the lack of information from local mining companies'. They expressed concern about mining companies not following the law at mine closures. They were also concerned about the Department of Mineral Resources and Energy (DMRE). One of them said that 'from a legislation point of view', there was 'a serious problem of non-compliance by both the DMRE and mining companies'. The official said that the DMRE, 'which supposedly should be leading the process', was not complying and was 'probably only informed by the mining companies once closures have occurred'.

Watson and Olalde (2019) echo the views of this respondent. They say that neither the regulatory requirements nor the disclosure norms are challenging or changing industry practice. Mines that close require a rehabilitation plan. But the Rustenburg municipality is excluded from obtaining information regarding mine closure. Consequently, most of our respondents believed that the mines were not complying with the regulations. The DMRE does not provide oversight, and both the mines and the DMRE bypass the municipality regarding mine closure because there are no legal requirements to interact with the municipality. But the mines' closure goals can increase municipal liabilities (an example of goal dependency). One municipal official said,

> The whole thing is chaotic. . . . They [the mining companies] would call us as local municipality to a meeting to tell us that they have (unilaterally) decided to donate certain portions of land and amenities that require much money for rehabilitation. . . . This is usually done without sharing any ideas or plans on how they intended to do rehabilitation. . . . We inherit without either capacity or plan to rehabilitate.

The views expressed by the respondents make it clear that there is no common approach to mine closure. The municipality, the DMRE and the mining companies operate in silos. Municipal officials said collaborative planning was lacking, despite the alignment of the municipality's IDP and the mines' SLPs (social and labour plans), which should promote such collaboration. We heard comments to the effect that the mines 'do their own thing and we do ours'. Respondents were also critical of the untransparent role of the DMRE. They felt that mine closure requires an inclusive process and dialogue (see World Bank & IFC, 2002; ICMM, 2008; World Bank, 2020). The failure to create a platform for collaborative planning on mine closures confirms that, despite 'early signs' or 'warnings', closure planning is usually delayed or not done. Attention is paid to the problems only when a mine closes (Macdonald et al., 2006).

Local companies and municipalities (Rustenburg Local Municipality included) tend to consider the consequences of mine downscaling and closure on an ad hoc basis. Despite many plans, Rustenburg has yet to succeed in diversifying its economy (Ntema, 2019; Crous et al., 2020; Marais, Campbell et al., 2021; RLM, 2021).

13.6 Mitigation measures

As part of the city's long-term plan to mitigate the devastating economic impact of the closures, the 2020/2021 IDP document refers to a new developmental path called 'Rustenburg beyond mining' (RLM, 2021). Research refers to such approaches as 'life-after-mining' (Maritz et al., 2021). The goal is to reduce the city's dependence on the mining economy through diversification. But we saw no evidence of the 'Rustenburg beyond mining' plan being put into practice. Mining remains the most extensive local economic sector, contributing 76.6% to the municipality's gross value added (RLM, 2022).

Despite the widespread consensus and acknowledgement of the urgent need to prepare for a post-mining Rustenburg, we found complacency and a reluctance or inability to implement such a plan. The 2022/2023 IDP document is evidence of this. It says 'there are no other sectors except for the mining sector with a comparative advantage . . . all other sectors have a disadvantage' (RLM, 2022, p. 43). Terms like 'mine closure' and 'mine decline' are absent, or omitted, from the document (RLM, 2022). This suggests that the municipality has difficulty acknowledging reality, and it is evidence of a cognitive lock-in. A senior municipal official said 'the impact of mine closures on our local economy is huge . . . we also see some key businesses leaving Rustenburg while fewer companies and businesses are showing interest to invest here'. But none of the municipal officials we interviewed mentioned 'diversification' as one of the municipality's possible interventions or countermeasures.

Several other factors complicate mine closure. For instance, the city's population increased from 534,000 in 2010 to 719,000 in 2020, with a projected figure of about 782,000 in 2025 (RLM, 2022). This translates to 3.03% annual growth, higher than the district and national annual averages of 2.42% and 1.59%, respectively. This growth is happening despite economic pressure from a shrinking mining sector. The city is thus struggling with an increasing backlog in the provision of basic service infrastructure, such as water, sanitation and electricity, because of the ever-growing number of informal settlements (RLM, 2021). In 2020, the city had about 39,400 informal dwellings (RLM, 2022).

13.7 Conclusion

Our Rustenburg case study shows how the mining companies, the municipality and the community have responded, or failed to respond, to closure. Overall, there is little transparency from either the DMRE or the mining companies. The result is that the municipality is left to deal with the consequences. Its ability to influence outcomes is limited, despite the principle of collaborative planning. We found no evidence that SLPS and IDPs are instrumental in managing decline. Although local governance should drive measures to deal with mine closures, it is unreasonable to expect municipalities to bear the full brunt of the decline.

The case study points to social disruption associated with mine decline, aggravated by the fact that mine decline has not led to negative population growth. Informal settlements are growing, and mafia-type operations and crime are common.

The goal of mine closure is to create dependencies for the municipality. For example, when the mines donate land to the municipality, they think they are providing an asset, but it is more often a liability. Areas in economic decline need less land rather than more to manage. Despite much talk about economic diversification, progress has yet to be made in building an economy beyond mining.

On the bright side, the relative success of the RBNDT may serve as an impetus to other community-based natural resource management initiatives in planning for a future without mining. The same cannot be said of other local efforts to benefit the Rustenburg community. Long-term plans remain unimplemented, and the focus does not go beyond the plan itself. A municipality that anticipated mine closure would have predicted a decline in municipal revenue and made long-term plans to deal with the situation. The case study points to the ease with which, in its IDPs, the municipality talks about economic diversification while ignoring those aspects over which it has control, such as planning regulations and municipal finance. Part of the problem is the planning system, which emphasises plan-making rather than implementation. Rustenburg Local Municipality currently lacks the capability to integrate the consequences of mine closure into all aspects of the functioning of the urban system.

References

Comaroff, J. & Comaroff, J., 2009. *Ethnicity Inc.* Chicago: University of Chicago Press.

Cook, S.E., 2013. Community management of mineral resources: The case of the Royal Bafokeng Nation. *Journal of the Southern African Institute of Mining and Metallurgy*, 113(1), pp. 61–66.

Crous, C., Owen, J.R., Marais, L., Khanyile, S. & Kemp, D., 2020. Public disclosure of mine closures by listed South African mining companies. *Corporate Social Responsibility and Environmental Management*, 28(3), pp. 1032–1042.

De Lange, A., 2019. The unintended consequences and spatial implications of mining downscaling in South Africa – A case of mining in Merafong City Local Municipality. Master's thesis, University of the Free State, Bloemfontein. https://scholar.ufs.ac.za/bitstream/handle/11660/10284/DeLangeA.pdf?sequence=1&isAllowed=y

ICMM (International Council for Mining and Metals), 2008. *Planning for Integrated Mine Closure: Toolkit.* London: ICMM.

Kgajane, I., 2022. Rustenburg mining community marches against unemployment and lack of service delivery. *SABC News*. www.youtube.com/watch?v=TCBoNdXaYUE&list=PPSV

Lagercrantz, H. & Khabbaz, S., 2019. The platinum boom in Rustenburg and the bust of the community: A case study of the natural resource curse in South Africa. Bachelor thesis, Linnaeus University, Sweden. https://lnu.diva-portal.org/smash/get/diva2:1342757/FULLTEXT02.pdf

Macdonald, C., McGuire, G. & Weston, H., 2006. *Integrated Closure Planning Review: Literature Review*. London: International Council on Mining and Metals (ICMM).

Marais, L., Campbell, M., Denoon-Stevens, S. & Van Rooyen, D. (eds.), 2021. *Mining and Community in the South African Platinum Belt: A Decade after Marikana*. New York: Nova Science.

Marais, L. & De Lange, A., 2021. Anticipating and planning for mine closure in South Africa. *Futures*, 125, p. 102669.

Marais, L., Ntema, J., Cloete, J. & Lenka, M., 2021. Mine housing, assets and informality in Rustenburg: Implications of mine closure. In Marais, L., Campbell, M., Denoon-Stevens, S. & Van Rooyen, D. (eds.), *Mining and Community in the South African Platinum Belt: A Decade after Marikana*. New York: Nova Science.

Maritz, J., Oranje, M., Van Huyssteen, E. & Nel, V., 2021. Planning for a post-mining future in the Rustenburg region and the transition towards it? In Marais, L., Campbell, M.,

Denoon-Stevens, S. & Van Rooyen, D. (eds.), *Mining and Community in the South African Platinum Belt: A Decade after Marikana*. New York: Nova Science.
Matebesi, S.Z., 2020. *Social Licensing and Mining in South Africa*. New York: Routledge.
Mayes, R., 2014. Mining and (sustainable) local communities: Transforming Ravensthorpe, Western Australia. In Brueckner, M., Durey, A., Mayes, R. & Pforr, C. (eds.), *Resource Curse or Cure? On the Sustainability of Development in Western Australia*. Heidelberg: Springer, pp. 223–237.
Mbenga, B. & Manson, A., 2010. *"People of the Dew": A History of the Bafokeng of Phokeng-Rustenburg Region, South Africa, from Early Times to 2000*. Johannesburg: Jacana.
Moerane, B., 2020. North West municipalities owed billions for rates and taxes. *SABC News*, 5 March. www.sabcnews.com/sabcnews/north-west-municipalities-owed-billions-for-rates-and-taxes/
Molotlegi, K.L., 2009. RBN review: State of the RBN address. *Opening of the Supreme Council, Royal Bafokeng Nation*. www.rbnoperationsroom.com/home/static/en_US/id/513/title/rbn+review+speech+annual+state+of+the+rbn+address.html
Mosiane, N., 2021. Livelihoods, the body, and the space of Phokeng, Rustenburg. In Marais, L., Campbell, M., Denoon-Stevens, S. & Van Rooyen, D. (eds.), *Mining and Community in the South African Platinum Belt: A Decade after Marikana*. New York: Nova Science, pp. 39–70.
Ntema, J., 2019. Rustenburg: Boom and bust in a mining town. In Marais, L. & Nel, V. (eds.), *Space and Planning in Secondary Cities: Reflections from South Africa*. Bloemfontein: Sun Media, pp. 203–217.
Oranje, M., Nel, V., Van Huyssteen, E. & Maritz, J., 2021. A brief history of platinum mining with a focus on the Rustenburg region. In Marais, L., Campbell, M., Denoon-Stevens, S. & Van Rooyen, D. (eds.), *Mining and Community in the South African Platinum Belt: A Decade after Marikana*. New York: Nova Science.
RLM (Rustenburg Local Municipality), 2021. *IDP Review: Financial Year 2021/2022*. Rustenburg: Rustenburg Local Municipality.
RLM (Rustenburg Local Municipality), 2022. *Draft Five Year IDP – 2022/2027*. Rustenburg: Rustenburg Local Municipality.
Royal Bafokeng Nation Operations Room, 2023. *Who are the Royal Bafokeng Nation?* www.rbnoperationsroom.com/home/static/en_US/id/6/title/who+are+the+royal+bafokeng+nation.html
Tau, P., 2020. "Rustenburg is not a place of hope any more" – Community devastated by massive job cuts in mining industry. *City Press*, 12 February. www.news24.com/city-press/news/rustenburg-is-not-a-place-of-hope-any-more-community-devastated-by-massive-job-cuts-in-mining-industry-20200212
Van der Watt, P. & Marais, L., 2021. Implementing social and labour plans in South Africa: Reflections on collaborative planning in the extractive industry. *Resources Policy*, 71, p. 101984.
Vivoda, V., Kemp, D. & Owen, J., 2019. Regulating the social aspects of mine closure in three Australian states. *Journal of Energy and Natural Resources Law*, 37(4), pp. 405–424.
Watson, A. & Olalde, M. 2019. The state of mine closure in South Africa – What the numbers say. *Journal of the Southern African Institute of Mining and Metallurgy*, 11(7), pp. 639–645.
World Bank, 2020. *Coal Mine Closure and a Just Transition for All*. Washington, DC: World Bank.
World Bank & IFC (International Finance Corporation), 2002. *It's Not Over when It's Over: Mine Closure Around the World*. Washington, DC: World Bank Group's Mining Department, Global Mining.

14 Rustenburg

Coping despite mine decline

Ngaka Mosiane and Cathy M Dzerefos

14.1 Introduction

Although two platinum mines were already in operation in Rustenburg in 1929, mining only became prominent in the early 1970s. Between 1993 and 2007, platinum mining became the dominant sector of the local economy. Rustenburg was known as one of South Africa's fastest-growing cities (Marais et al., 2021). During the boom, Rustenburg attracted migrant workers looking for jobs. Although the city's economy generally slowed down in 2007, mining remained dominant while the socio-economic environment deteriorated. This chapter investigates the disconnect between the mining companies and the community and the effect this has had on the community. We argue that despite losing out on Rustenburg's economic growth and being badly affected by mine decline, ordinary people have shown remarkable resilience in adapting to the changes in the local economy.

To investigate the trajectory of mining developments in the western platinum belt and their implications for the lives of ordinary people, in 2019 we interviewed 63 ordinary people (not involved in mining) living in Freedom Park (42 women and 21 men), an informal settlement near Rustenburg (see Figure 14.1). We wanted to find out how they generated their social and economic spaces to rework or even overcome their marginality and dependencies. The chapter is based on the accounts of four of the respondents, one man and three women. Some of the respondents in this study were members of the Freedom Park Development Forum and the Freedom Park Housing Committee.

Chapter 13 describes in detail the decline in Rustenburg, and we will not repeat this here.

The global financial crisis of 2008/2009 was followed by the Marikana Massacre of 2012 and prolonged strikes at the mines near Rustenburg in 2014. However, in 2020/2021, platinum performed well despite the COVID-19 pandemic and brought some stability to the area during that difficult time. The area experiences continual boom and bust cycles, but the general trend for the community is one of decline.

The people of Freedom Park are migrants and immigrants from many parts of South Africa, southern Africa and Asia. About 4,000 government-subsidised houses, known as RDP houses,[1] were built in this neighbourhood in 2003, but about 4,835 people continue to live in informal housing structures (StatsSA, 2011). In

DOI: 10.324/9781003403326-14

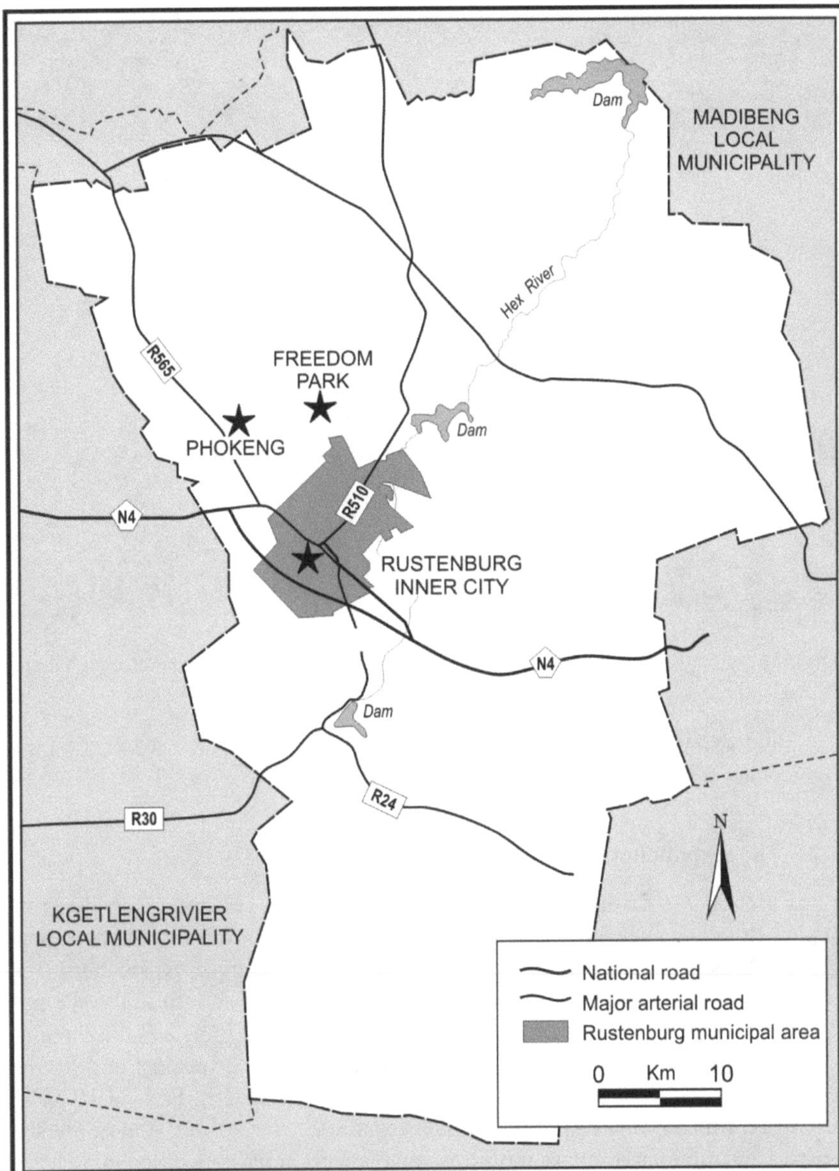

Figure 14.1 Freedom Park in Rustenburg Local Municipality

addition to the living-out allowance or residence allowance, mining houses promote home ownership. Accordingly, in 2013, Impala Platinum Company (Implats) officially opened Platinum Village, an ostensibly affordable development of just over 2,000 new freestanding bonded houses (Implats, 2013). Implats provided a soft loan for mineworkers (to increase the monthly living-out allowance) of, for

Figure 14.2 Freedom Park informal settlements, RDP housing areas and Platinum Village

example, R480,000 for a three-bedroom house (R340,000 bond + R140,000 interest-free loan from) (see Figure 14.2; Implats, 2013).

14.2 The respondents' stories

In this section, we mainly use the narratives of the four selected respondents but supplement, where needed, the other 59 interviews. The stories represent ordinary people's responses to the workings of the mining economy in Rustenburg. Figure 14.3 shows a typical informal house in Freedom Park,[2] an informal settlement established in the 1980s by migrants looking for work at the Impala Platinum mine.

Monghadi was one of the first immigrants to work as a resident and hawker in Freedom Park. He left Johannesburg and came to Freedom Park in 1984. He anticipated finding business opportunities around the Rustenburg mines. He sold second-hand clothing, military uniforms and blankets at the gate of the mine hostel number 8. Up until the end of the 1990s, he made a worthwhile monthly income. Miners' paydays were especially good, and he profited from their bonus payments and 'skaters' (the end-of-day cash pay on holidays). He also built a tuck shop to sell goods in the community. He occasionally served as a church warden. He managed to support his children at institutions of higher learning. In the conversations with Monghadi's children, they acknowledged his role in their education. Some of his profits were invested in classic cars. He recounted the joy of driving his 1984 Mercedes Benz 380 SL or his 1975 Ford Fairlane, saying that 'everyone looks at

Figure 14.3 A typical housing structure in Freedom Park in the late 1980s and the 1990s

you, and claps their hands and those who are not jealous will wish you well' and that at traffic lights, 'fellow drivers will commend you for what you are driving'. He also owned an RDP house, which he did not occupy as it was too small for him but handed on to other family members.

Other immigrants and migrants – informal traders, job seekers and, later, miners – followed; squatting increased on nearby Bafokeng land in the 1980s (see also Chapter 14). When chased by the police, they frustrated the police's efforts by escaping to nearby private land. In search of permanent residence, some migrants attempted to acquire plots in Bafokeng villages. Some were accepted as members of the Bafokeng tribe; others were not. Nokwanda said she was one of those who were denied residency when she joined her late husband, a miner, in 1977. As a result, Freedom Park became an alternative base from which her family eked out a living. She augmented her husband's salary by selling beer, cooked food and soft drinks at the gate of hostel number 8. She also made an income by renting out 12 shacks for R50 each per month. Adding this to her husband's salary, she was able to pay for her three children's schooling and later for vocational and tertiary institutions. The children are now independent and self-sufficient, one of whom is a qualified teacher in Freedom Park. Nokwanda also receives a government old-age pension and lives in an RDP house she inherited from her husband after he died. Her daughter's teaching salary helped her expand the RDP house into an impressive seven-room house with a satellite television dish and an Internet connection. Figure 14.4 shows an example of such a house.

Broadly, Monghadi's and Nokwanda's stories indicate the significant benefits that ordinary people's livelihood activities yielded in the 1980s, linked to increased

Figure 14.4 An RDP house turned into a modern commodity

profitability in the mining sector. Their livelihood activities were fewer yet stronger and more certain than they are today, combining a limited portfolio of any of the following activities and assets: informal trading, renting out shacks, old-age pension, a scholarship and a range of social support networks (an extended family and collective political resistance in Freedom Park, for example) (later, a subsidised house would be added). Nokwanda's children grew up living alternately in the Eastern Cape with relatives and in Rustenburg with their parents. Similarly, Monghadi's children lived with their parents and relatives in Rustenburg, Johannesburg and Lesotho. Three of Monghadi's children have qualified in engineering from Further Education and Training.

Beyond the goal of giving their children a leg-up, Monghadi's and Nokwanda's stories also show that ordinary people participate in livelihood formation to pursue aspirations (status, for example) beyond mere survival. Their histories are examples of incorporation into Western modernity, albeit on its fault lines and fissures (Mosiane, 2021, 2023).

Monghadi's Mercedes Benz and Ford Fairlane were, for him, a symbol of status. The same could be said about Nokwanda turning an RDP house into a fancy, plastered house connected to the world through a television satellite dish and the Internet. After the interview at her house, she expressed concern about the unwillingness of her neighbours to improve the aesthetics of their surroundings. She and Monghadi aspired to achieve their goals of personal and public distinction as they rented out backyard rooms, sold beer and cooked food, and sold second-hand goods and soft drinks. Their children had moved up in the world, away from informal trading into teaching and engineering.

The livelihoods and profitability of the informal activities of the ordinary people of Freedom Park can be traced to the increased militancy of the labour force in the mid-1980s and the concomitant wage increases. Higher wages meant that more money was available for spending in the informal economy. Freedom Park became a space for livelihood formation for migrants, the unemployed and miners. This shanty settlement was demarcated into blocks, A to F, each with its own committee. The Block Committees arbitrated for the defaulters of informal loan and rent repayments. They also mediated in family disputes involving financial claims by women when a cohabitation relationship broke up or by legal wives coming from rural areas to check on husbands who had not visited home for a long time. The people of Freedom Park also established a self-defence unit to intervene in broader community disputes, mainly making up for the absent state (there being no police station in Freedom Park at that time and no basic services currently). The unit patrolled the settlement at night and ensured safety and control. It also arrested residents suspected of stealing livestock and cars from the surrounding Bafokeng villages or stealing power cables from the local mines. It also took on the self-appointed role of enforcing values, such as attempting to stop people from having multiple sexual partners and ensuring that women dressed 'properly'.

A new wave of migrants arrived in the Rustenburg area in the 1990s as the mining economy was growing and the political landscape was changing nationally. However, the growing mining economy was also increasingly outsourcing[3] and mechanising its work processes. This restructuring of work processes and the increasing local population threw Freedom Park into transition as a place of livelihood formation. The opportunities to earn a living in Freedom Park were diminishing as more workers were laid off and the local circulation of cash receded. Although the residents were poorer than their predecessors, they remained resilient, also aiming to pursue other personal goals beyond just survival. Their livelihood activities were more diverse and unreliable than those of their forbearers, as they combined more numerous portfolios of livelihood activities and assets. With that said, mobility remained a constant for this generation of the 1990s, as was the case with the immigrants and migrants of the 1980s. Mobility is so fundamental to their lives that it is space itself (Malaquais, 2007). That is, ordinary people spend so much of their time travelling that their home spaces extend to distant areas. Teboho, for example, left her home in Lesotho in 1990 to find work to support herself and her family back home:

> In Johannesburg, I lived with a married miner from Lesotho, who helped me pay rent and buy food. Since I could not get enough money from him to send back home, I moved to a farm in the Mpumalanga Province. There I lived with another Sotho man who used to divide his salary between his family and me, giving me R900 a month and taking R1,000 to his family. I lived with him until 1997, when I moved elsewhere in Mpumalanga because he had to leave his job to care for his family's livestock in Lesotho. Although he continued sending me R500 a month afterwards, I decided to move to Freedom Park in 1999, at which time he also stopped giving me money. I am now living with an unmarried Tswana boyfriend. I used to sell beer to augment the money he gave me, but I stopped doing that because I did not make money anymore.

Teboho moved from one country to another, from villages to farms, and ultimately to a city, to find a place to make a living. Her words reflect the realities of many women in Freedom Park.

Abongile's story is similar. In 1980, she left her troubled home in Taung in the North West Province and moved to the town of Welkom in the Free State Province. She moved from one place to another for 15 years: to a farm in Welkom, back to the town of Welkom, to Durban, back to Taung (though she did not visit her home), to Klerksdorp in the North West Province, to KwaZulu-Natal, to Klerksdorp again, to Carletonville in Johannesburg, back to Taung (living with her parents for some time) and finally to Rustenburg in 1995, where she found Freedom Park, a place where she could make a home and a living. The 15 years of being on the move were a struggle. At each stop, in desperate need of food and accommodation, she unwillingly 'moved in' with a man. She was subjected to sexual and physical abuse. Her 'boyfriends' beat her, and she sometimes hit them back in self-defence. In Welkom in the late 1980s, she once hid in a storeroom to avoid police arrest for working in 'South Africa', outside her Bophuthatswana 'homeland' without a permit.[4] After being told that her troubles signified an ancestral calling, she responded by training as a traditional health practitioner in KwaZulu-Natal, but she dropped out because of the lengthy training and the large amount of domestic work she had to do.

Abongile, Teboho and many women like them sought stability to rebuild their lives in very specific ways: helping their children through the education system beyond high school (which was clearly beyond their financial ability), building modern houses (in some cases elaborate ones), accumulating decorative furniture, setting up businesses, participating in the public life of Freedom Park and creating some room for privacy. On arrival in Freedom Park, Abongile found a footing by cohabitating with a married man who gave her R500 a month, from which she supported her family and participated in stokvels.[5] In the shack settlement in Freedom Park, where Abongile lived at first, a person's life and livelihood assets (or lack thereof) were largely collective experiences. Women living in shacks supported each other financially and psychologically. Much activity went on outdoors because the shacks were so small, and this made for sociable living. Abongile also participated actively in the life of Freedom Park, volunteering as a member of the Policing Forum when Freedom Park did not have a police station. This work paid some money to volunteers, which she used to buy some furniture and appliances while still living in the shack settlement (a kitchen unit, a stove, a fridge and a videocassette recorder). She said, 'I buy these things to show what I am working for. I grew up liking nice things'. In the early 2000s, she received an RDP house, which she hoped would be an asset for her children in case of her death.

Notably, and in line with Simmel's ideas of indifference and blasé attitudes (Levine, 1971), the RDP house gave her some privacy that she did not have when she lived in the shack settlement. The introduction of a police station in Freedom Park also meant that she no longer had to share her problems with members of the community (neighbours, the Block Committee or the Policing Forum). 'In the past', she said, 'I used to share my problems with my neighbours. Now I have some privacy due to my RDP house, and I pray when I have problems'. The acquisition of satellite television and services such as water and electricity, all inside the RDP house, allowed her to enjoy some private living, even when times were hard. An RDP house accelerates commodities and consumer desire; possessions contribute to defining social relationships. People's consumption lifestyles enable them to stand out from their neighbours and be recognised. The RDP housing programme thus, to some extent, eroded neighbourly connections and social structures like the Policing Forum. At the same time, Freedom Park has incorporated rural ways of living. It has become a jumble of urban and rural lifestyles.

Abongile rebuilt her life, thanks to a small income from community volunteer work, financial assistance from her boyfriend and her son's contributions to food and clothing for his siblings. It was also because of these income sources that she could commit to one man. Women with insufficient financial support from a partner may enter into sexual relationships with about five men at a time. Women support each other in this, tactically using each other's shack rooms for sex work.[6] Implats, a mining company, recruited several of them to work as 'peer educators' on HIV/AIDS, sexually transmitted diseases and the use of contraceptives, paying them R200 a month. The National Department of Health later took over the project and increased their pay to R1,000 a month.

Since 2003, the Cape Town-based organisation Sonke Gender Justice has been working with them, flying them to Cape Town for training on issues of sexual and reproductive health and their human rights and security as sex workers. They thus feel empowered to return to their home villages to run similar peer education projects. The work of the Catholic Diocese of Rustenburg in Freedom Park, discussed in the following, has had similar effects on the lives of another group of women who work as caregivers to HIV/AIDS patients in Freedom Park and Rustenburg. It has helped them to transform their lives under conditions that are not of their choosing, especially the economic downturn brought about by mine decline.

Some of the women take pride in using their RDP plots as a kind of rural space, keeping goats and planting crops, in keeping with what they can afford and the lifestyle they grew up with. Other women, and men, tap into African traditional spiritual worldviews (Hay & Dzerefos, 2022) that can help them make a living in a community affected by mine decline. They claim to offer spiritual services such as healing rituals to help people get married, conceive a child or find employment, or to protect them from bad luck and witchcraft. During cleansing rituals, the person seeking solace may symbolically leave behind in a city river of natural areas such as Magaliesberg Mountains an item that represents their stress such as clothing or an employer's warning letter. Others may be advised to collect holy water from or leave consumable offerings at sacred sites in Magaliesberg (Dzerefos, 2021). It is common to see a household preparing the early morning meal offerings to their ancestors, thereby bringing the hope that they and their families will be protected from harm and helped to realise their dreams. These practices give people an anchor for facing life's challenges.

Many of the 1990s' generation of the people of Freedom Park are surviving or even flourishing by seeking opportunities, building on the groundwork of their 1980s predecessors, and obtaining networks of support from the city of Rustenburg and beyond. The Freedom Park Housing Committee delivered, through the government social housing scheme, more than 4,000 houses. Basic services such as electricity, water and sanitation were partially provided to housing recipients. Freedom Park has generally seen some level of infrastructural development since the year 2000, and such projects have largely been possible because of the community's ability to forge links with actors in many places through the support of the Rustenburg Catholic Diocese, which has a long tradition of working with the indigent in Rustenburg.

The Diocese helps to empower neighbourhoods neglected or yet to be reached by the state. Through funds from local mining companies, notably Amplats, Implats, Lonmin, Aquarius and Xstrata, the Diocese often builds temporary structures to serve as schools, health centres and sites for income-generating community projects. It worked with the Freedom Park Development Forum to source temporary structures for a school, a crèche, a police station and a private clinic. It also secured some R200,000 from the British High Commission, Impala and the local association of medical doctors for community-based projects such as a makeshift bakery, fence making, car battery charging and garment sewing. When such projects are registered and operate as community and state entities, the church leaves the area and moves on to another needy one. This is how the Catholic Diocese came to work in Freedom Park in 1997.

The people of Freedom Park partnered with the Catholic Diocese as a point of contact for support from local, national and international state and private institutions. Thus, at the level of institutional support (faith-based agencies and private organisations and companies), some transformative relationships across and beyond the city were forged, with the potential to improve the lives of the urban poor. Such institutional support is necessary for attaining social goals in a situation of mine decline, like in Rustenburg. This support and the associated projects are, however, often unsustainable. The bakery and sewing projects in Freedom Park are just getting by, and the rest of the projects have had to close their operations after the manager defrauded the projects' bank account and disappeared. The overall development process in Freedom Park has consequently taken on a more charity-oriented trajectory. The Catholic Diocese initially adopted an approach that empowered people through community-based projects. From around 2003, its programmes narrowed down to HIV/AIDS treatment. Alongside the Orphans and Vulnerable Children (OVC) programme, it focused on the Anti-retroviral Treatment Programme, an outreach programme rendering home-based healthcare, nursing, counselling and support services. Crucially, although the local state is well-positioned to act as a guarantor of social and economic development in both distributional and empowerment senses, it could have been more present in many other initiatives. Only a quarter of the people living in Freedom Park had received formal housing from the National Housing Department by the mid-2000s.

14.3 Reflections on mine decline and closure

The aforementioned narratives reflect on community responses in an area strongly affected by the ups and downs of mining. The narratives point to the value of migration for ensuring livelihoods. Under apartheid, the government monitored migration in detail, although the narratives provide evidence of people defying these rules. The post-apartheid dispensation brought freedom to decide where to settle and how (or whether) to invest in housing. In the same way that mobility enabled people to move to Rustenburg, mobility can be an asset that enables them to leave when a mine closes. People like Abongile, who spent 15 years on the move, will have experience that could serve them well if the economy of Rustenburg declines to a point where they can no longer make a living and they have to leave. On the other hand, ownership of a house, at first considered a blessing, could become a handicap. Overall, the post-apartheid housing policy does not consider the value of migration and overemphasises permanency and stability. It suffers from goal dependency, being firmly set on countering historical migration processes (Marais, 2023). Alternatively, the role of the state in improving mobility could also entail seamless sub-regional road integration. Road connectivity would ease mobility for people whose valuable assets are sunk in Rustenburg to remain living locally but working elsewhere in the broader Gauteng city region. As regards the local economy, seamless road integration can improve the value of the adjacent land and stimulate local enterprise resulting from the passing trade.

Although some of the informality described earlier is problematic, informality does have some advantages when mines decline. Issues of permanence and impermanence are central to living near a mine. Not having a long-term fixed investment in housing enables mobility. Despite various plans to develop a post-mining economy for the area, the likelihood of these plans bearing fruit is slim. The state-provided housing helps create the permanence necessary for development. Our respondents talked about using the government-provided house as business premises and investing in it by extending it at their own expense. However, the evidence also shows that many residents do not want this permanence. One of the respondents was concerned that her neighbours were not upgrading their houses. She wanted permanence, but not everybody sought it. Some saw few incentives to invest in an area whose long-term viability was unclear. Renting a shack at a small cost and being free to move was a way to deal with the uncertainty of mining. Mining economies like Rustenburg illustrate both the permanent and impermanent aspects of housing, and understanding the pros and cons of both is important in the context of mine closure.

The narratives reflect the social disruption that mining brings. Respondents spoke of gender-based violence and crime. Despite the expectation that mine closure in Rustenburg will result in an outflow of people, a large percentage may remain in the area. Social disruption is likely to continue and increase because of the decline. Many of the case studies in this book come to this conclusion, and Rustenburg is likely to be no different. The existing community governance structure may stop functioning, and the loss of jobs and income will cause social stress. Mine investments in social support will disappear. The narratives reflected on projects that had helped them in difficult times. Mine closure will lead to these projects being discontinued, despite the need for continued social support.

There are many examples of how people take over governance functions. The Freedom Park community did this. The respondents told us how the community governance actors provided order through land-use planning, social services and protection (performing the role of the police). This governance had some negative implications. For example, controlling how women dress impedes personal freedom and could be perceived as gender bias (they did not tell men how to dress). Such rules were a response to rapid growth and the absence of a government. Predicting how mine closure will affect this current situation is difficult, but community governance may decline. Leaders may move on first, leaving a void. This void, for example the community losing its ability to police the area, could change the scale and nature of social disruption during the decline.

The study found evidence that interdependence between the mines and the community is diminishing and will be completely broken when mining companies leave. Full employment on the mines has shifted largely to contract work. Clean wages have probably increased the miners' monthly take-home salaries. However, they are still vulnerable to retrenchments at closure, and backup funds to deal with the immediate consequences of closure may be inadequate. Mining companies will withdraw from social spending in the area. The likelihood of long-term social disruption is high. The poor housing conditions of many Freedom Park residents are also closely associated with the goal dependency associated with miners' housing. Leaving the hostels did not improve housing conditions for most people. Instead, it placed further pressure on the government to provide adequate housing. The initial hostile approaches in the 1980s to informal settlements by the Bophuthatswana police stand in contrast to the mushrooming of informal settlements since 1994.

However, it also appears that some Freedom Park residents have succeeded in diversifying their incomes and accumulating assets. We heard of people buying cars, renting out shacks, extending the house they lived in, opening a shop, having multiple income sources and educating their children. People who have managed like this despite living in an area of mine decline will help to diminish the effects of mine closure in the long run.

14.4 Conclusion

This chapter complements the other case study of Rustenburg (Chapter 13), which criticises the way the local government has responded to the likelihood of mine closure. This chapter relates the stories of some ordinary people living in the peripheries of the city, surviving or even flourishing. It shows how they responded to the mining economy – some just getting by while others are making a decent living in Freedom Park.

The uncontrolled growth and social disruption associated with the mining busts will continue to have repercussions when the mines close. The absence of local government in optimising opportunities during the years of plenty could have further negative implications with mine closure. For example, the lack of reliable, potable water has been a problem in both formal and informal settlements, and residents require income to buy it (Staff Reporter, 2023). In a sluggish economic climate, buying water would become increasingly problematic for many people, and environmental health problems such as cholera outbreaks from drinking polluted water or death from dehydration during a heat wave could increase. Overall, a far more nuanced approach to governing mine settlements is required. The absence of government is not helpful, and the current focus on housing and the drive for homeownership could be detrimental at closure.

Much of the social disruption created by the mining decline will continue, despite a population decrease. Dependence on the platinum mines for employment is likely to diminish. Finding alternative mining or other work will be difficult. This trend will extend to those indirectly dependent on mining. The absence of social programmes from the mine because of the closure will cause further hardship.

However, a note of hope comes from the aforementioned narratives, which illustrate the resilience of ordinary people despite the challenges of mining volatility. The emphasis in mining studies is often on miners. Our chapter shows how ordinary people indirectly make a living from mining. We can only speculate about how mine closure might affect them, but the aforementioned narratives suggest resilience. We further documented intergenerational value in the way their efforts have given their children access to higher education. These children's qualifications will enable them to find better jobs than their parents had, thus helping to diversify household livelihoods and minimise the long-term consequences of mine closure.

Notes

1 These are houses built under the government's Reconstruction and Development Programme.
www.dhs.gov.za/content/government-subsidised-houses
2 Originally called 'Wildebeesfontein', but we use the name 'Freedom Park' in this chapter for simplicity.
3 Outsourcing of work refers to short-term employment with reimbursement restricted to the service provided. This has become an acceptable and growing practice in South Africa from the level of a household garden service to the mining sector. Outsourcing of work allows employers the freedom to hire and fire without being accountable to labour legislation or unions (Forrest, 2015). It negates employees benefits related to housing and pensions, and undermines trade unions' ability to organise and negotiate (Chinguno, 2013).
4 Homelands were created as national ethnic states for black people and migration between these homelands and South Africa was strictly regulated.
5 A stokvel is a club that serves as a saving scheme in South Africa. Members contribute money to a central fund and a different member receives the collected money each month.
6 At the time of writing, sex work was illegal in South Africa. A draft bill to decriminalise it was published for comment in December 2022.

References

Chinguno, C., 2013. *Marikana and the Post-Apartheid Workplace Order*. University of the Witwatersrand: Society Work and Development Institute. https://bit.ly/2q2yqJ4

Dzerefos, C.M., 2021. A participatory approach to developing environmental literacy to protect the freshwater resources used by African Apostolic faith communities. *The Conservation Symposium, Pietermaritzburg*. 1–5 November. www.youtube.com/watch?v=PzUGFJZdyqg

Forrest, K. 2015. Rustenburg's labour recruitment regime: shifts and new meanings. *Review of African Political Economy*, 42(146), pp. 508–525.

Hay, A. & Dzerefos, C.M., 2022. Recognising the sacredness of nature to promote education for sustainable development. *Indilinga: African Journal of Indigenous Knowledge Systems*, 21(1), pp. 39–53. https://rb.gy/nzuvmg

Implats, (2013). *Impala Accommodation Strategy*. [Online]. https://implats.co.za/pdf/presentations/2013/impala-accomodation-strategy-5-december-2013.pdf, accessed on 10 June 2021.

Levine, D. (ed.), 1971. *Simmel: On Individuality and Social Forms*. Chicago: University of Chicago Press.

Malaquais, D., 2007. Douala/Johannesburg/New York: Cityscapes imagined. In Murray, M.J. & Myers, G.A. (eds), *Cities in contemporary Africa*. London: Palgrave Macmillan.

Marais, L., 2023. *The Social Impacts of Mine Closure in South Africa: Housing Policy and Place Attachment*. London: Routledge.

Marais, L., Campbell, M., Denoon-Stevens, S. & Van Rooyen, D., 2021. *Mining and Community in the South African Platinum Belt: A Decade after Marikana*. New York: Nova Science.

Mosiane, N., 2021. Livelihoods, the body and the space of Phokeng, Rustenburg. In Marais, L., Campbell, M., Denoon-Stevens, S. & Van Rooyen, D. (eds.). *Mining and Community in the South African Platinum Belt: A Decade after Marikana*. New York: Nova Science.

Mosiane, N., 2023. Remaking of the historical Rustenburg landscape. In Mosiane, N., Murray, J., Peberdy, S., Sithagu, A. & Dzerefos, M.C. (eds.), *Landscapes of Peripheral and Displaced Urbanisms: GCRO Research Report*. Johannesburg: Gauteng City-Region Observatory.

Staff Reporter, 2023. Water crisis in Rustenburg persists. *Platinum Weekly*. www.platinumweekly.co.za/article.php?id=7452&categoryID=8

StatsSA, 2011. *Census 2011 Community Profiles*. Pretoria: Statistics South Africa. www.statssa.gov.za/

15 A framework for understanding the social aspects of mine closure in South Africa

Lochner Marais, Sethulego Matebesi and Verna Nel

15.1 Introduction

This book investigates the local social consequences of mine closure. We wanted to understand how mine closure affects places, communities and their organisations and how they respond. The book's first part investigates theoretical concepts and policies regarding mine closure. The second part consists of ten case studies (one on retrenched miners and the other nine place-based). We were concerned about the fragmented nature of the mine closure literature and wanted to bring together evidence from mining sites across South Africa, including some that are in remote areas. In this final chapter, we take a comparative approach to the case studies.

Through the case studies, the book reflects on four mining sectors: diamonds, gold, coal and platinum. The varying trajectories of these four commodities have different local consequences. Although this book is on South Africa, it has a wider application: the liberalisation of the mining industry in Africa over the past three decades has made it more urgent to think carefully about the effects of mine closure. We hope this book will provide a foundation for further work on the continent.

We started the book with an anecdote about a general manager of a mine who had made his own personal plans for a post-mining future but was concerned about what the mine's suppliers would do. He feared they would not be able to remain concerned. When the mine closes, their businesses will close too.

This is the book's essence: how do we plan during the life of a mine for the day when it ends?

Mines usually do this, but public planners and small businesses, like the suppliers, often respond inadequately, too late or not at all. South Africa has a draft strategy for mine closure (DMRE, 2021, see Chapter 3). But, as we have outlined in Chapter 3, there are still areas for improvement and much need for simplification of complex and obstructive bureaucratic procedures. This final chapter is an attempt to broaden the ideas regarding mine closure. We point to the complexities and dependencies often ignored in the draft mine closure strategy.

The arguments in this chapter might be misread as ignoring the importance of economic diversification. This is not our intention. We recognise the importance of economic diversification, but it is only one avenue to explore. Our concern with the emphasis on economic diversification is four-fold: it is portrayed as the only

DOI: 10.324/9781003403326-15

option, it seldom takes into account economic realities, it does not view decline as a potential economic diversification strategy and it ignores what ordinary people do.

Instead of focusing obsessively on economic diversification, governors should think about what decline means for infrastructure, land use and the social environment.

15.2 Theoretical contribution

One of our reasons for writing this book was the dearth of theoretical thinking about mine closure. A decade ago, Digby (2012) proposed framing mine closure as a process of deindustrialisation. This has value. We find the concept of social disruption and the dependencies identified by evolutionary governance theory useful for identifying commonalities in the disparate findings of our case studies. Social disruption is a useful concept to help us understand and engage with the current practice of social and environmental impact assessments. Historically, these assessments have been to investigate the possible consequences of mine development. In this book, we use the concept of social disruption to investigate the consequences of mine closure. Evolutionary governance theory makes local and other decision-making more central in understanding responses to decline and closure. It provides a conceptual framework to explain why planning and implementing alternative economic options is difficult after mine closure.

15.2.1 Social disruption

Social impact assessments use the term 'social disruption'. We explain in Chapter 2 how this concept evolved through phases of social impact assessments in boomtowns (mostly in the mining and energy contexts). The large investments usually result in an influx of people who disrupt local norms and values. Yet Smith et al. (2001) showed that the disruption declines over time. The question is whether closure results in new forms of social disruption, such as increased crime. Some evidence points to the opposite. For example, O'Connor and Ruddell (2021) argued from a case study in North America that mine closure leads to a decrease in crime. The outflow of people from the mining area was the main reason. However, in our case studies, mine closure does not always lead to an outflow of people. In some cases, an outflow of people coincided with more land invasions (expanding the municipal land area). Many people remain behind in these economically deprived areas and suffer social consequences.

The case studies point to several examples of social disruption following mine closure for individuals and the community. Most case studies reported high unemployment levels. Social disruption was evident in crime (including organised crime), environmental destruction, gender-based violence, divorce and children not completing school. The disruption of education for the second generation is worrisome. Chapter 5 described the consequences for households of losing a job in the mining sector. Typical problems were loss of health insurance, disruption of education and having to relocate. Although similar consequences follow the loss of a non-mining job, retrenchment is a bigger blow in the mining industry because it is so hard for a miner to find an alternative job.

These social problems are also related to weak municipal governance in South Africa's mining areas. The Koffiefontein case study shows how difficult it is to maintain physical infrastructure in the context of decline. A municipality's inability to provide and maintain infrastructure worsens the social disruption caused by job losses.

In Chapter 1, we used Vivoda et al.'s framework (2019) to show the wide range of social aspects of mine closure and commented on the value of this framework for understanding mine closure and its consequences. We have expanded this framework, as shown in Table 15.1. We have added two more social aspects to the list: local government finance and informal (artisanal) mining (a form of path dependency, which we discuss in the next section). We split two of Vivoda et al.'s categories: 'security and education and training infrastructure' becomes 'security' and 'education and training', and 'housing' and 'health', previously one category, become two. We add three columns for evidence from the book, implications for thinking about the social impacts of closure, and indications of dependencies.

We think that despite the value of Vivoda et al.'s framework, it is simply a list of a set of possible social aspects associated with mine closure, with no value judgement or distinction of positive and negative effects. We need a deeper understanding of the issues. In our expanded version, we have attempted to take an integrated approach to cause and effect (not just provide a list), to highlight the possible negative consequences, to integrate the relationships between issues in the list, to consider the dependencies associated with mining and mine closure and to indicate potential individual or community trauma.

Social impact assessments are important for large infrastructure investments. These assessments explain cause and effect and integrate the relationships between different aspects. Most social impact assessments directly or indirectly refer to positive and negative impacts. For example, the influx of people will lead to an increase in crime. In our framework, we have tried to show this by adding an implications column, and we have provided ample evidence of these relationships in the discussion that follows.

Evolutionary governance theory points out the dependencies between mining and mining communities. We identify some of these in the framework. The discussion about these dependencies that follows provides further clarity. Evolutionary governance theory and the related dependencies add two important aspects to understanding the implications of mine closure: a governance angle (how communities and their structures make decisions) and how planning for growth could make coping with decline much more difficult. This should be understood in the context of historical and current decision-making for managing growth.

15.2.2 Dependencies

We use the dependencies associated with evolutionary governance theory as the framework for this discussion (see Chapter 2).

Table 15.1 Social aspects of mine closure – an expanded framework

Social aspect	*Indicative elements*	*Evidence in this book*	*Implications for thinking about the social impacts of closure*	*Indications of dependencies*
Economy	Local economic activity (diversity and dependence)	Much evidence of economic decline, yet inadequate response. Decline often leads to economic diversity but does not eliminate the dominance of the mining sector. Emalahleni: Despite planned closure, no evidence of local planning for it	• None of the areas in the case studies in this book had developed an alternative economy beyond mining. This does not mean it is impossible but the show how hard it is. This reality should counter idealistic plans. • Alternative responses to mining often fail. • Households have less, or no, income and declining living standards bring social problems.	• Path dependency in the inability to find new economic paths. • Planners ignore path dependency in the existing space economy. • Cognitive lock-in in the form of ignoring decline, not seeing existing economic changes and over-idealistic plans (no recognition that these plans need to be funded).
	Household income	Qualitative evidence of lower incomes		
	Local living standards	Qualitative evidence of lower incomes hence lower living standards		
Business	Local business development opportunities in the mining sector	Local business opportunities are disappearing. In Tshikondeni, Kleinzee, Koffiefontein: reduced access to goods and products for local people	• Little evidence that former miners create successful new businesses • Obtaining shares in a mining company as a form of compensation had no benefits (Richtersveld).	• Typical examples of the breaking of interdependencies between mining and other sectors of the economy.
	Business opportunities in other sections	Much evidence of business closures in other sectors to loss of clients or shrinking market		
Employment	Local employment opportunities in the mining sector	Most case studies reported declining employment in mining.	• Losing a job in the mining industry means it is difficult to find alternative employment. • Miners unprepared to deal with losing a job (sometimes despite help from the mine) • Often have inappropriate skills for alternative employment	• Evidence of path dependency because of cognitive lock-in. Inability to consider the implications of losing a job in the mining sector
	Local employment opportunities in other sectors	Most case studies reported declining employment in non-mining sectors.		
	Local employment stability/volatility	Some evidence of declining mining and increasing non-mining employment (Rustenburg). Generally, a decline or expected further decline		

Security	Social order and safety (tensions, violence and crime)	Many case studies provide evidence of increased crime, gender-based violence and social disruption. Richtersveld: continued conflict between the mine and the community	• This is a major concern because it shows that mine closure, like mining growth, causes long term social disruption.	• Often, informal mining continues long after mine closure (path dependency).
Education and training	Local skills development in mining and other sectors	We found little evidence of this, apart from in the Tshikondeni case study. Chapter 5 shows disruption of children's education because of parents' job loss in mining (intergenerational effect).	• Those with skills move out first.	
	Access to quality education and training	Kleinzee: school closures and how that inhibits the development of new economies Tshikondeni: closure of ECD centre		
Infrastructure	Local transport (buses, roads and airports) Critical infrastructure (for food, power, water, telecommunications)	Matjhabeng: closure of the. Plans to reopen it Koffiefontein and Kleinzee: difficulty maintaining infrastructure after closure.	Evidence of infrastructure decline and vandalism because of closure	• Closure creates a new path: deteriorating infrastructure and reduced access to infrastructure spur further social problems. • Development of open towns and normalisation – some evidence of goal dependency in this respect

(*Continued*)

Table 15.1 (Continued)

Social aspect	*Indicative elements*	*Evidence in this book*	*Implications for thinking about the social impacts of closure*	*Indications of dependencies*
Amenity	Local aesthetic and recreational resources (heritage sites, parks, recreation areas, communal areas)	Koffiefontein, Kleinzee, Matjhabeng: municipal services have declined.	• Further decline in access to social infrastructure is likely.	• Often, this decline results from the drive to normalise towns or develop open towns – goal dependency.
	Local culture, arts and sports (including facilities)	Matjhabeng: losing its sports prominence in the Free State Province		
Livelihoods	Local livelihoods (access to land, food, water and shelter that affects livelihoods)	Most case studies reflect on how mine closure has affected people's livelihoods. Chapter 15 provides some good examples of the potential indirect impacts for non-miners.	• Social impact assessments should think about both direct and indirect consequences of closure, including possible benefits.	.
Land	Local land access, ownership and use	Mines donate land to municipalities	• Long-term consequences of communities getting mining assets that become liabilities	• Donation of mining land to municipalities is an example of goal dependency
	Recognition of traditional, customary ownership	Richtersveld: an attempt to use mining as compensation for a historical land claim		
Housing	Local housing quality, availability and affordability	Major emphasis on housing as a strategy to bring stability. Closure implications not considered	• Need to consider tenure types that suit the temporary nature of mining	• Housing policies focusing on ownership are examples of goal dependency. They have long-term implications because of the objectives the government set.
Health	Community health and wellbeing Access to quality health and social services	Miners losing access to medical benefits of employment, clinics closing (Tshikondeni, Kleinzee). Emalahleni: closure could lead to better health outcomes by reducing pollution	• Disrupts the healthcare needs of mineworkers who lose their jobs • Mine closure could have positive effects for health.	

Environment	Environmental aspects that affect social conditions (quality of air, water, land, ecosystems)	Emalahleni: closure could reduce pollution and water use. Tshikondeni: closure puts new pressure on the environment. West Rand: radiation, acid mine water, unrehabilitated land	• Although not the main focus of this book, mine closure in areas like Emalahleni would mean an immediate improvement in air quality.	• Environmental consequences result from path dependency on mining. and will continue long after mine closure.
Demography	Local population dynamics (growth/decline, migration, ageing, gender balance)	Case studies' evidence varies: population decline, slower growth and continued growth despite lack of prospects	• Decline seldom means a rapid decline in population. • Often, populations keep growing, although the growth may be slower than in other areas.	• Dependence on the goal of providing land and housing in mining areas. Closure can, paradoxically, bring population increase and land invasions.
Participation	Stakeholder participation in closure planning, and closure and post-closure processes (including decision-making)	Very little evidence of participation in planning for the decline No evidence of dovetailing SLPs and IDPs		
Inclusion	Inclusive stakeholder engagement with vulnerable and marginalised groups (Indigenous peoples, women, ethnic minorities, disabled, elderly, young) in closure planning, and closure and post-closure processes (including decision-making)	Very little evidence of participation in planning for the decline No evidence of dovetailing SLPs and IDPs		

(Continued)

Table 15.1 (Continued)

Social aspect	*Indicative elements*	*Evidence in this book*	*Implications for thinking about the social impacts of closure*	*Indications of dependencies*
Social (general)	General socio-economic considerations	Social services diminishing or unable to deal with the problems		
	Social considerations in financial assurance mechanisms	No evidence of this. Mining companies reduce social spending as closure sets in.		• Social spending can become a goal dependency ignoring the reality of decline and closure.
Aspects added from this book				
Local government finance	Reduction in income from rates and taxes	Rapid decline in municipal income	• Municipalities struggle to remain a going concern.	• Decline in local government finance is an example of broken interdependency between mines and municipalities.
Informal (artisanal) mining	In many cases, formal mining is replaced by informal mining.	Matjhabeng, West Rand, Richtersveld: informal mining as a result of mines not being formally closed	• Positive local economic implications but negative social consequences	• Informal mining is evidence of a path dependency – continued dependence on mining.

Source: Expansion of Vivoda et al.'s framework (2019, p. 8)

15.2.2.1 Interdependency

Mining towns depend on mining, and mining depends on these towns and their governance. Many of the case studies were in rural communities (Richtersveld, Tshikondeni), small towns (Kleinzee, Koffiefontein) or secondary cities (Emalahleni, Matjhabeng, Rustenburg, West Rand), where this kind of dependency is prominent. Interdependency can be positive or negative. The towns depend on the mines for economic activity, and the mines depend on the towns to provide a good living environment for their employees. This interdependency is positive when it creates livelihoods but negative when social disruption occurs. Mine closure disrupts this interdependency, ends it or creates a new one.

Mining growth usually leads to an influx of people. This influx to support the mine's labour needs also supports local economic development. The decline is expected to lead to an outflow, breaking the interdependency between the mine and the town. The evidence in the book provides evidence of both. In Kleinzee, the town's economy declined rapidly, leading to the outmigration of people at closure. The same pattern was visible in Tshikondeni. In other cases, there is only evidence of the slowing down of growth rates but not a numerical decline in people. In many places, mine decline puts pressure on the municipality to make land available. For example, some former miners invade the land, increasing the pressure on municipalities to provide infrastructure and increasing the municipal urban areas through urban sprawl (see Chapter 6 on the West Rand and Chapter 13 on Rustenburg).

A declining mining economy has interdependencies with other economic sectors and the social environment. Losing one's job in mining is problematic because finding another job in another sector is difficult (see Chapter 5 on retrenched miners and Chapter 11 on Tshikondeni). Interdependency between ordinary people and the mines was evident in the second Rustenburg case study (Chapter 14). Many people make an indirect living from mining by selling goods and services to mine employees. They find it difficult when mines close and the market for their products declines. Mine closure or decline might break many interdependencies. The implications for ordinary people dependent on mining can be severe.

Many case studies found that retail enterprises had closed, severely inconveniencing the local population (Chapter 8 on Kleinzee, Chapter 9 on Koffiefontein and Chapter 11 on Tshikondeni). This infrastructure was usually set up to ensure that the mine would attract a quality workforce, satisfying both parties. The mine closure breaks this interdependency. These case studies record how people travel further (at a higher cost) to access basic services. Schools and clinics have closed, with detrimental local consequences. Schools and other public services are central to economic development in small towns and rural communities. Losing these functions because of mine closure has ripple effects that prevent post-mining economies' development.

Many municipalities develop long-term interdependencies with mining in three main areas: municipal finance, infrastructure and land use. Municipal finance is dependent on collecting local rates and taxes from landowners. Mine closure negatively affects the collection of rates and taxes (Chapter 6 on the West Rand,

Chapter 7 on Matjhabeng and Chapter 13 on Rustenburg). In some cases, mines and municipalities share the same infrastructure. In Emalahleni, this extends to the mining companies providing water services to the municipality. Municipalities are expected to provide land for fast-growing mining towns. They usually fund this through external finance from another sphere of government. But the benefits of this cease at closure. Repaying a loan to finance new land development for mining might be hampered by the mine closing. Infrastructure and service interdependencies hold long-term risks. The example from Koffiefontein (Chapter 9) shows how difficult it is to maintain infrastructure without the mine helping and how municipalities need to take over these services when mines close. In turn, the inability to maintain infrastructure can cause social problems.

Interdependencies tend to be short-term and can lead to path dependencies. For example, while a mine is in operation, the company depends on the miners for labour, and the miners depend on the company for income and other benefits. When the mine closes, the interdependency is broken. The company no longer needs the miners, but they still need the company. The stronger entity in this coupling, the company, can walk away undamaged, but the miners, and their community, have been set on a path of dependence on the mine and may find themselves destitute.

All the interdependencies the chapters identify – between mines, municipalities and community – all depending on each other are positive in that they enable livelihoods. It is the breaking of these interdependencies, when the mine withdraws, that is responsible for the situation turning negative and causing social disruption. The old company town set-up could be seen as a positive interdependency, working well for all the parties involved. The miners and their households had the benefits provided by the company, and the company had a contented workforce. However, the negative effects appear when the mine closes, the interdependency is broken and the miners struggle to embark on a new path of independent living.

When mines close, another interdependency, between communities and municipalities (the former paying for services, the latter receiving income), is broken. Municipalities can no longer provide services after losing their biggest client, the mine. Interdependency no longer works but becomes a vicious circle: a community that cannot or will not pay for services, and municipalities that cannot provide services unless the community pays.

15.2.2.2 Path dependency

The case studies in the book provide many examples of what can result from path dependency. Clinging to mining as the only possible path means that towns cannot diversify their economies, former miners cannot find alternative employment and informal mining continues, with all the problems it brings. Path dependency can result from cognitive lock-in (discussed later in this section). At the end of this section, we discuss several material dependencies as examples of path dependencies.

Mining makes finding alternative economic paths difficult. None of the case study areas has been able to diversify its economy (with the exception of Mogale City in the West Rand, which shows some promise). Despite the decline and mine closure and the elusiveness of economic diversification, mining remains a dominant economic sector in all but one of the case study areas (Tshikondeni, where the mine has closed). Other reasons for this are the nature of the mining industry, public planning processes and remote locations.

Mining creates specific patterns in the space economies of countries. They become an economic path from which it is difficult to diverge. Although most mining occurs in remote areas, the main services to the industry (legal and financial) occur in gateway cities where no mining occurs (Atienza et al., 2021). In these remote areas, no alternative economic sectors are available to cushion the impact of mine closure because the country's broader economic benefits seldom filter through to the remote mining areas but accrue to the more central non-mining towns. It is difficult to break this mining-induced pattern; thus, many places are stuck in the path dependency that mining creates. Local procurement programmes, where mines buy goods and services (e.g. security or cleaning services) from suppliers in their own area, are an attempt. But at best, they have short-term benefits that disappear when the mine goes, and they can create other long-term dependencies.

Diversifying away from mining is difficult because it is difficult to deviate from a path that was once taken, especially mining. Mine developments often represent radical economic change in spaces with slow economic change (usually remote locations). Mining usually changes the socioeconomic environment of a remote location, creates growth and increases employment (although sometimes not for the local people). This happens because of large mining upfront investments, which are a front-end approach to economic development. These investments are usually substantially larger than non-mining investments. Miners, public governors and planners think the same approach will work for economic diversification. One large capital investment will change the economic landscape. This type of thinking makes three invalid assumptions. It assumes that amounts similar to those for mining investments are available and that the state can make these investments; that there is no need to consider the existing local value chains and logistics or that the investment, as with mining investment, will be big enough to counter these constraints; and that local skills and capacities are available to drive the new economy. Because these assumptions are false, the dependence on the mining path continues.

These assumptions easily spill over to the public response. Some problems are examples of cognitive lock-in, but others ignore the economic lock-ins that mining creates, which are the result of the realities of the space economy in remote areas. Evolutionary economic geography emphasises the importance of space and time and the slowness of economic development. Despite mining usually being an exception to slow change, planners seldom understand the temporal and spatial constraints of remote mining areas. Competing in fast-moving goods is difficult because of the remoteness of the locations. Niche products might fare better, but they too must overcome spatial constraints.

The focus on steering in the planning system is another kind of path dependency (see Chapter 7 on Matjhabeng). Marais et al. (2021) have argued that basing local planning on the new public management paradigm inhibits economic diversification for one main reason: it de-links planning and implementation as the emphasis is on planning (steering) and not implementation (rowing) (see Chapter 2). The absence of implementers (the private sector or network governors) in steering hinders efforts to find an alternative to mining.

The chapters are replete with the results of sticking to the path of mining in the face of mine decline. For example, the case study on Kleinzee shows how mine closure has led to crime, which hampers the town's tourism potential. At a different level, organised crime in declining gold mining does the same (West Rand and Matjhabeng). For example, syndicates strip iron and other metals from all new infrastructure.

Another form of path dependency is in the nature of mining jobs: once set on the path of mining, most other paths are closed to you. Chapter 5 (a study of retrenched mine workers) and Chapter 11 (Tshikondeni) showed how difficult it is to find non-mining employment after retrenchment. The concept of relatedness is central to this problem: economic diversification occurs when industries are related to existing ones. This means building on a region's existing skills. The nature of mining skills makes it difficult to find alternative jobs because few related sectors are available. This is the case even though mining jobs are becoming more mechanised.

Cognitive lock-in is a common reason for path dependency. The owners of the Tshikondeni mine were transparent about the closure. They communicated the imminent closure to their employees, the local community leadership and the municipality. Other mines have not been so transparent. But the diverse responses after closure show how difficult it is to talk to people about mine closure. Some do not want to hear the message, and it is difficult to understand a negative message at a time when the mine is prospering. The case studies provide examples of three kinds of cognitive lock-in: denial, belief that it is only an economic problem, and belief that mining will confer lasting prosperity.

People tend to believe that the mines will be there forever. The Koffiefontein case study shows that it was impossible to get the local and provincial governments to take closure seriously. The expected transition in the coalfields of Mpumalanga (Chapter 12 on Emalahleni) is a further example. Despite much talk, actions remain absent. There is evidence of cognitive lock-in in the Richtersveld and Tshikondeni case studies. In the Richtersveld, the community accepted shareholding in a state-owned mining company as compensation for a land claim. The expectation of riches inhibited thinking about the likelihood of a decline and the company not paying dividends.

The assumption that closure is only an economic problem has the logical consequence that the solution is economic diversification. Planners pay little attention to planning the other implications of decline (land use, land-use regulations, infrastructure). There was little evidence in any of the case studies of planning for decline in municipal finance, right-sizing or the ecological restoration of the town.

Mining companies can be guilty of encouraging cognitive lock-in by giving the impression that they confer lasting value. They talk about 'creating value beyond closure'. Anglo American (2023) talks about 'ensuring a lasting legacy' for mining regions. Do not get us wrong. The fact that mining companies are considering aspects of closure is commendable. Our concern is the overemphasis on their value.

In many cases, this is a marketing campaign rather than a realistic approach to closure. The rhetoric is unhelpful for two reasons: implementation is difficult, and there is an implicit assumption that the mine remains responsible for the region's prosperity.

Informal mining, which often replaces formal mining, is an example of path dependency. People in mining towns struggle to find any livelihood path other than mining. Failure to close mines, or closing them inadequately, enables informal or illegal mining. Although informal/artisanal mining stimulates local economies, it has negative social consequences. Marais et al. (2022) found higher crime levels in areas of mine decline than in most other areas of South Africa.

Finally, there are material dependencies. We list them here as examples of path dependency because they hamper the development of new economic paths. Unrehabilitated mining land, people living between mine dumps, acid mine drainage, the radiation hazard on the West Rand, and new housing and infrastructure in places with declining economies are physical factors that represent an established path inhibiting the development of new economies.

15.2.2.3 Goal dependency

Several goal dependencies are identifiable from our case studies. We identify the main ones here: government dependence on the goals of growth and creating modernity through mining, homeownership to make reparations for apartheid's migrant labour system and local procurement.

New mining developments place pressure on towns. They must provide infrastructure and housing for a fast-growing population. Most local governments need help to do this. Then, in the wink of an eye, the same people are told to plan for decline and closure. But the government remains focused on the goal of growth, modernity and an end to migrant labour. It is difficult to change gears from planning for growth to planning for decline. We later recommend that social impact assessment extend to mine closure. This requires thinking about mine closure at the start of the mining process. Most mines do this. Yet this is only sometimes done by public planners. Governors and planners should see growth and decline not as opposites but as a continuum.

The South African housing approach in mining communities is another example of goal dependency (Marais, 2023). The push for homeownership and planning for new stands or plots does not consider closure. This goal is largely driven by the Department of Human Settlements, but has implications at the local level. Municipalities must deal with larger pieces of land during growth but with lower income from taxes when closure sets in. Expanding settlements simply because a government department views them as a way to create stability causes problems at closure. To some extent, open-town and normalisation policies do the same.

The government's goal dependency – its stubborn pursuit of a goal without considering negative consequences – is evident in the drive for local procurement; the admirable goal of local benefit could have adverse consequences. Our book started with an anecdote drawing attention to the problems suppliers will have when a mine closes. Although local procurement systems have some value, they must be considered carefully, as mine closure will remove the suppliers' market. Similar concerns can be expressed about the social support programmes of mines. It is easy to direct these social programmes to deal with growth concerns and never consider the implications of decline.

Mine closure disrupts these goals, and the dreams are shattered, leaving local people suffering long-term consequences.

15.3 Policy and planning considerations

15.3.1 The mine life cycle and planning

Mine life cycles are becoming shorter because of mining efficiencies, technological advances and the increased dominance of surface mining (although this varies across commodities). Booms and busts are often short-lived. These frequent cycles tend to disguise the overall trend of decline. Shorter mine life cycles have local implications. The period of local benefit might be shorter, reducing the time available to respond to mine closure. In many cases, the volatility will increase, making planning difficult. Van Assche et al. (2020) offer ideas on the potential of land use tools to mitigate the effects of these cycles on communities.

The many links between growth and decline require a conceptual adjustment. Instead of seeing these as different phases, they should be viewed as a continuum. This has implications for social impact assessments, which focus mainly on the initial investment. These assessments do not consider decline. Anglo-American's closure toolbox is a move in the right direction (Anglo-American, 2019), but it still sees the start and end of a mine as different processes.

The chapters reveal many examples of plans for economic diversification. But these public plans have yet to follow a life-cycle approach. There are two reasons for this state of affairs.

First, mining companies seldom share their intelligence, assumptions or risks. None of our case studies showed collaboration regarding business intelligence. We do not see appropriate communication between mining companies and municipalities. Consequently, planners in the public domain seldom have access to this business intelligence. Mining companies argue that such information could make countering their business strategies easier for their competitors. Although there is some disclosure of mine closures in sustainability reporting (Crous et al., 2020), this reporting only happens near the closure date.

And second, planners do not apply life cycle planning. The absence of appropriate intelligence is central to this problem, but cognitive lock-in is also part of the problem. Life-cycle planning is not considered if mining closure is unthinkable or blurred by visions of 'the good times'.

The international literature on shrinking cities has emphasised the value of decline (Pallagst et al., 2014). Valuing decline provides space for ecological restoration and a basis for economic diversification. Accepting decline can lay the foundation for the development of new economies. The land rehabilitation of mining sites is important, but it should also apply to the land rehabilitation of residential areas. Rather than developing grand plans for economic diversification, planners could make economic diversification possible by understanding shrinking and the responses of local people. Chapter 15 (based on fieldwork in Rustenburg) tells the stories of ordinary people in the mining context and considers the consequences of decline for them. Appropriate policy responses require careful consideration of what people do in response to decline.

Seeing growth and decline as a continuum should make it clearer how to design infrastructure for settlements in a mining town. At least two possibilities exist: develop or upgrade infrastructure so it can take higher densities, and consider the potential value of modular infrastructure. Higher densities reduce urban sprawl, which could reduce long-term land management liabilities. This should go hand-in-hand with the promotion of rental housing, an appropriate form of tenure for temporary economic activities, because it enables mobility (Marais, 2023).

Modular infrastructure is commonly used in information technology environments. It is less common and more difficult to use in civil environments, although renewable energy systems use it. Modular infrastructure would make it easier to navigate growth and decline as it is easy to add or remove modules, or to dismantle them and move them elsewhere. However, the main value of using modular infrastructure for a mining town is that it would be a response at the beginning of mining that would help address concerns at closure.

15.3.2 Finding appropriate planning responses

The case studies illustrate a range of ways that planners deal with mine closure: doing nothing, making wildly over-ambitious plans, ignoring existing examples of economic diversification and failing to consider that municipalities have some control through their land use management system. All four are examples of cognitive lock-in.

Planners suffering from cognitive lock-in assume that mine closure will not occur. Or they assume the future will look glamorous and prosperous like the mining days. They ignore the difficulty of finding economic alternatives in remote locations. They ignore economic diversification that already exists in their town, and they seldom consider how land-use planning and regulation play, or could play, a role in economic diversification. This is especially true in the larger places (Emalahleni, Matjhabeng, Rustenburg, West Rand). For example, the expansion of the regional services function in Matjhabeng does not feature in its local economic development plan. Neither do plans consider removing red tape in the land-use management system if decline becomes a reality. Growing areas require high levels of regulation. Areas that are experiencing economic decline need more flexibility and less regulation. Yet these ideas are seldom part of planning and scenario-building.

Compounding the problem of cognitive lock-in, the time frames of the planning system are largely inadequate. IDPs have a typical horizon of five years. Aligning that five-year cycle with the mine's SLP does not support long-term planning, especially not planning for closure (if that is 20 years down the line). Although spatial development frameworks (SDFs) have longer time frames, the mining and local planning processes remain largely based on five-year plans as required by the legislation.

The planning system needs reconsideration. Marais et al. (2021) have argued that the planning system seems to have drawn its ideas from the new public management paradigm that emphasises plan-making and steering (as opposed to implementation). The consequences of plan-making and steering were visible in most of the case studies in the book. Most plans were unrealistic, and we found no evidence of implementation. Many thought that having a plan solved the problem. The emphasis on plan-making results in ignoring resource constraints and implementation processes. For example, the local economic development plan in Matjhabeng assumed that other spheres of government would fund all the large infrastructure projects. It did not matter that provincial and national governments are unlikely to make these investments in an economic climate where government finances are under pressure. Neither did it matter that the municipality could not contribute financially to the plan.

Cognitive lock-in, inappropriate time frames and constraints in the planning system hamper mine closure planning. The question is, what alternatives are there? We think consideration should be given to the following ideas: The literature on shrinking cities has taught us that it is possible to plan for decline. However, this idea is foreign to the South African planning system. Planning decline requires understanding the existing dependencies between mines and local governments.

A planning system should consider the long-term implications when planning mining investments. We have already pointed out that booms and busts should be viewed as a continuum. What local governments do during a mining boom has implications for closure. Some closure implications can be prevented during growth. We mention options such as promoting rental housing, high densities and modular (temporary) units. Policies of homeownership and extensive land development have long-term implications when mine closure occurs.

South African mining policy uses the terms 'mine closure' and 'ghost town' in the same breath. Government policy states that ghost towns must be avoided. But nobody considers the likelihood that one can plan for a ghost town, revegetate or rewild the area, forget about it and move on.

It is often overlooked that mine closure could have positive implications. In Emalahleni, coal mining and coal-fired power stations pollute the air, with negative health implications for the local population. Closure would bring cleaner air and reduce health problems. It would also save water, as the power stations are big water users. More water would be available for other sectors, like agriculture or the production of green hydrogen.

Our case studies show that local governments have very little understanding of what they can do to plan for mining busts. They make utopian plans but ignore what is within their capability. We found plans detached from reality and pipe dreams requiring substantial inter-governmental funding. Sometimes the mines are responsible for these utopian ideas. Mining-dependent municipalities should work much harder to understand mining processes. Some practical steps they could take are:

- Accept that mine closure is an eventual reality. Mine closure is not only the result of resource depletion. It is often the result of price volatility or reduced demand for the resource.
- Hold regular discussions with mining companies about their current problems and future scenarios.
- Develop a variety of economic scenarios.
- Trace the prices of commodities over the long term. The mining industry releases papers on the prospects of a specific commodity. Mining-dependent municipalities should access these papers and determine the implications for their locations.
- Develop a register of risks and map out dependencies. This would include a critical review of all government actions in mining areas.
- Integrate these risks into possible future scenarios. Once these scenarios are in place, determine the implications for the local economy, land use, infrastructure provision and the social environment.
- Communicate these risks in clear and simple ways to the community, but show how the municipality wants to mitigate them. Communities need to plan for decline and change.

15.3.3 Rethinking the open town policy

Despite the value of open towns and normalisation, mine closure creates problems for open towns. Normalised and open towns are largely geared towards growing mining economies. They broaden the participation of people in public management and are designed to integrate mining and non-mining economies. They are a way to ensure that mining brings public benefit. Since the start of the democratic dispensation in 1994, South Africa has embarked on normalisation and mainly created open towns. Most of the case studies in this book are of open towns. Alexander Bay in the Richtersveld is in the process of being normalised. Kleinzee has been normalised. Koffiefontein is still partly a company town, as the mine owns a substantial portion of the houses.

Mine closure brings to light problems with normalisation. Public entities now become responsible for managing both booms and busts. A study of Emalahleni by Van der Watt and Marais (2019) shows that the capacity to cope with this responsibility is seldom available beyond metropolitan areas, and the result is infrastructure decay and social problems.

Delinking the town and the mine means that the town is managed without the business intelligence of the mine, yet planners need help understanding mining trends and the likelihood of closure. Not knowing these trends or the level of available reserves can lead to overoptimistic economic scenarios. Many of the case studies of open towns in this book are examples of this problem.

Normalisation usually leads to either resettlement or transferring the title of the houses from companies to individuals. Many households, therefore, receive a housing asset. Van der Watt and Marais (2019) found mixed outcomes from this process. On the one hand, there was evidence of households building assets, valuing homeownership and becoming more independent. On the other hand, not everybody wanted homeownership, and the poor public management of services has affected the market negatively. The policy of promoting homeownership in mining towns needs to consider the importance of migration for miners who lose their jobs. Normalisation allows mining companies to shift the closure responsibility and the problems to local governments and households.

Our call here is not to change all open towns back to company towns. Rather, we argue that promoting normalisation requires a case-by-case approach. In some cases, the company town model is appropriate. Normalising all company towns as a matter of principle, as we have seen in South Africa since 1994, is an ill-considered response. Where mines are in a remote area, where there is little likelihood of economic diversification away from mining or where mining is on a small scale, there is good reason for company towns to prevail. Consideration must be given to whether an open town can attract competent municipal staff to manage it. Remoteness might make this difficult.

15.3.4 Integrating environmental and social aspects of mine closure

Environmental aspects of mine closure dominate mine closure literature. Rehabilitating mined land does indeed have environmental benefits, and revegetation is important. But the socio-economic benefits of rehabilitating mining land must be considered too. Rehabilitation, both surface and underground, could prevent illegal mining. This is especially important in gold mining, and several chapters reflect on the social consequences of illegal mining by syndicates. Marais et al. (2022) found that crime levels in declining gold mining areas are much higher than elsewhere in the country. As well as preventing illegal mining, the rehabilitation of mining land makes it available for potential new uses. Mine dumps and tailings inhibit new post-mining activities. Level ground, free of acid mine water, can be a site for new economic opportunities. Land rehabilitation in former gold mining areas on the West Rand should help reduce radiation and improve living conditions. Impact assessments need to recognise that land rehabilitation is not only an environmental concern but also needs to be integrated with social concerns. This is especially important where settlements exist among mine dumps and tailings (Matjhabeng and West Rand).

15.3.5 SLPs are inappropriate for managing decline

We found very little evidence of collaboration between mines and municipalities. In theory, they dovetail their SLPs and IDPs, as they are required to do. This dovetailing seems to have some merit when mines are growing. Yet none of our case study authors reported the dovetailing of SLPS and IDPs in the declining towns they studied. We think there was nothing to report on. It simply does not work. There are several industry-related blockages:

- Mines plan for social funding at the start, not the end of their life cycle, so funds for social aspects are unavailable at the end.
- The legal focus on environmental aspects means that many social issues fall off the radar.
- Many mines go bankrupt or go into business rescue, resulting in insufficient funds for closure.
- Long technical delays on the government's side and uncertainty about time frames make proper closure planning difficult.

Among the reasons why linking SLPs and IDPs does not work for local government closure planning are:

- Too much emphasis is placed on planning for growth. Planning for decline and closure is a huge cognitive leap for local governance and planning.
- The original plan seldom makes provision for closure.
- The capacity to integrate closure planning into normal planning is lacking.

15.3.6 Planning modular towns, and planning for ghost towns

Planning for booms and busts takes work, technical capacity and acknowledging the reality of closure. Part of the problem is that many decisions about mine development and closure must be embedded in a local approach or response, taking into account both growth and decline. We have argued that growth and decline should be seen as a continuum and not as the opposite sides of a coin. At the same time, government policy documents insist we should not have ghost towns. We think planning a modular town is critical for a mining town, and it would be appropriate in some cases to plan for a ghost town.

The following would be the main ways to create a modular town:

- Start with the premise that mining is a temporary economic activity that will lead to temporary land uses.
- For electricity, plan mini-grids that link to renewable energy.
- Plan civil infrastructure with modular components that can be added to, replaced, or removed.
- Consider using new algae-managed sewerage plants, which are cheaper than conventional sanitation, but also consider other alternative sanitation systems.
- Plan high densities that reduce urban land cover.
- At the start of the mine's operation, consider possible alternative uses for mining land when mining ends.

To plan for a ghost town:

- Start with the premise that mining is a temporary economic activity that will inevitably end.
- Plan mini-grids for electricity.
- Build environment-friendly infrastructure that can be easily downscaled and integrated with the environment.
- Plan high densities to minimise land use.
- Using building materials that can be easily demolished or re-used.

15.4 Further research

This book addressed the main shortcomings of the mine closure literature: too little reflection on the social aspects, a literature produced mainly by the mining industry, single case studies and a lack of theoretical thinking. It provided a platform for a more comparative approach to understanding the social aspects of mine closure in various resource environments and remote areas. The book is not the last word. We do not provide simple answers. Our research conclusions are tentative. We hope the discussion we have started will extend into mining boardrooms and the offices of policymakers in Pretoria. We argue that rather than making fixed policy statements, South Africa must carefully consider local contexts and governance while recognising the temporary nature of mining.

We think four types of research could help take the mine closure research forward.

First, the theoretical work could be expanded. For example, although we provided ample evidence of interdependencies, these could be assessed in more detail and with statistical rigour. For example, how do new jobs in mining or the loss of jobs in mining affect jobs in other sectors of the economy? Such analysis would provide evidence of interdependency and help understand patterns of economic diversification (or the lack thereof).

Second, we would like to see a more concerted effort to consider the narratives of ordinary people, link them conceptually and determine the implications for social impact assessments. Chapter 5 (on retrenched miners) and Chapter 15 (narratives from non-mining people in Rustenburg) attempted to do this. But more could be done to find out how local people and mining households negotiate their way through boom and bust. The capabilities approach could help us understand their freedoms and agency. Local governance could benefit from their narratives. It should be possible to link planning and qualitative research approaches. This would have the value of revealing what people do to counter the risk of mine closure. Too often, the literature portrays them as helpless bystanders to policy and economic realities. To remedy this defect, the researchers who contributed the case studies for this book went out and talked to ordinary people.

Finally, a qualitative research approach involving the stories of people in mine communities could help show the value of migration processes as part of mine closure. We still do not know enough about how people use migration to mitigate the consequences of closure. Many government policies assume that migrant labour has come to an end. Research shows the opposite (Marais, 2023; Pretorius & Blaauw, 2022).

References

Anglo American, 2019. *Mine Closure Toolbox*. Version 3. www.angloamerican.com/~/media/Files/A/Anglo-American-Group/PLC/sustainability/mine-closure-toolbox-version-3-2019.pdf

Anglo American, 2023. *Ensuring a Lasting Legacy*. www.angloamerican.com/our-stories/environment/ensuring-a-lasting-legacy

Atienza, M., Arias-Loyola, M. & Phelps, N., 2021. Gateways or backdoors to development? Filtering mechanisms and territorial embeddedness in the Chilean copper GPN's urban system. *Growth and Change*, 52(1), pp. 88–110.

Crous, C., Owen, J., Marais, L., Khanyile, S. & Kemp, D., 2020. Public disclosure of mine closures by listed South African mining companies. *Corporate Social Investment and Environmental Management*, 28(3), pp. 1032–1042.

Digby, C., 2012. Mine closure through the 21st century looking glass. In Fourie, A. (ed.), *Proceedings of the 7th International Conference on Mine Closure*. Perth: Australian Centre for Geomechanics, pp. 33–38.

DMRE. 2021. *Draft Mine Closure Strategy*. Pretoria: DMRE.

Marais, L., 2023. *The Social Impacts of Mine Closure in South Africa: Housing Policy and Place Attachment*. London: Routledge.

Marais, L., Ndaguba, M., Mmbadi, E., Cloete, J. & Lenka, M., 2022. Mine closure, social disruption and crime in South Africa. *Geographical Journal*, 188(3), pp. 383–400.

Marais, L., Nel, V., Rani, K., Van Rooyen, D., Sesele, K., Van der Watt, P. & Du Plessis, L., 2021. Economic transitions in South Africa's secondary cities: Governing mine closures. *Politics and Governance*, 9(2), pp. 381–392

O'Connor, L. & Ruddell, R., 2021. After the downturn: Perceptions of crime and policing in the southeastern Saskatchewan oil patch. *The Canadian Geographer*, 5(3), pp. 281–291.

Pallagst, K., Wiechmann, T. & Martinez-Fernandez, C., 2014. *Shrinking Cities: International Perspectives and Policy Implications*. New York: Routledge.

Pretorius, A. & Blaauw, D., 2022. Residents' perceptions of coal mining and energy generation. In Marais, L., Burger, P., Campbell, M., Denoon-Stevens, S.P., & Van Rooyen, D. (eds.), *Coal and Energy in South Africa: Considering a Just Transition*. Edinburgh: Edinburgh University Press, pp. 135–149.

Smith, D., Krannich, R. & Hunter, L., 2001. Growth, decline, stability, and disruption: A longitudinal analysis of social well-being in four western Australian communities. *Rural Sociology*, 66(3), pp. 425–450.

Van Assche, K., Gruezmacher, M. & Deacon, L., 2020. Land use tools for tempering boom and bust: Strategy and capacity building. *Land Use Policy*, p. 103994.

Van der Watt, P. & Marais, L., 2019. Normalising mining company towns in Emalahleni, South Africa. *The Extractive Industries and Society*, 6(4), pp. 1205–1214.

Vivoda, V., Kemp, D. & Owen, J., 2019. Regulating the social aspects of mine closure in three Australian states. *Journal of Energy & Natural Resources Law*, 37(4), pp. 405–424.

Index

Note: Page numbers in *italics* indicate a figure and page numbers in **bold** indicate a table on the corresponding page.

FSC
www.fsc.org
MIX
Papier | Fördert gute Waldnutzung
FSC® C083411

Kolibri 360 GmbH
Ferdinand-Jühlke-Straße 7
99095 Erfurt, Deutschland
produktsicherheit@kolibri360.de